Digital Signal Processing Laboratory using MATLAB®

Sanjit K. Mitra
University of California at Santa Barbara

Boston   Burr Ridge, IL   Dubuque, IA   Madison, WI   New York   San Francisco   St. Louis
Bangkok   Bogotá   Caracas   Lisbon   London   Madrid
Mexico City   Milan   New Delhi   Seoul   Singapore   Sydney   Taipei   Toronto

WCB/McGraw-Hill

A Division of The McGraw·Hill Companies

6 7 8 9 BKM BKM   0 9 8 7 6 5 4 3 2

Manual ISBN 0-07-232246-2
Package ISBN 0-07-232876-2

http://www.mhhe.com

Contents

Preface

Digital signal processing (DSP) is concerned with the representation of signals as a sequence of numbers and the algorithmic operations carried out on the signals to extract specific information contained in them. In barely 30 years the field of digital signal processing has matured considerably due to the phenomenal growth in both research and applications, and almost every university is now offering at least one or more courses at the upper division and/or first-year graduate level on this subject. With the increasing availability of powerful personal computers and workstations at affordable prices, it has become easier to provide the student with a practical environment to verify the concepts and the algorithms learnt in a lecture course.

This book is for a computer-based DSP laboratory course that supplements a lecture course on the subject. It includes 11 laboratory exercises with each exercise containing a number of projects to be carried out on a computer. The total number of projects may be more than what can be completed in a quarter- or a semester-long course assuming a three-hour per week laboratory. It is suggested that the instructor select pertinent projects that are more relevant to the lecture course he/she is teaching. If the computer laboratory is open for longer hours, it is recommended that the student be encouraged to come to the laboratory for longer periods of time to enable him/her to complete all projects.

The programming language used in this book is MATLAB,[1] widely used for high-performance numerical computation and visualization. The book assumes that the reader has no background in MATLAB and teaches him/her through tested programs in the first half of the book the basics of this powerful language in solving important problems in signal processing. In the second half of the book the student is asked to write the necessary MATLAB programs to carry out the projects. I believe students learn the intricacies of problem solving with MATLAB faster by using tested, complete programs and later writing simple programs to solve specific problems. A short review of the key concepts and features of MATLAB is provided in Appendix A.

Altogether there are 75 MATLAB programs in the text that have been tested under version 5.2 of MATLAB and version 4.2 of the Signal Processing Toolbox. The programs listed in this book are not necessarily the fastest with regard to their execution speeds, nor are they the shortest. They have been written for maximum clarity without detailed explanations. This book includes a diskette containing all MATLAB programs for the PC

---

[1]MATLAB is a registered trademark of The Mathworks, Inc., 24 Prime Park Way, Natick, MA 01760-1500, phone: 508-647-7000, **http://www.mathworks.com**.

running Windows 95/98, the Macintosh PowerPC running Mac OS 7 or higher and UNIX workstations. All programs are also available via anonymous FTP from the Internet site **iplserv.ece.ucsb.edu** in the directory **/pub/mitra/Labs**.

Each laboratory exercise contains a number of projects for the student to implement on their computers. Each project is followed by a series of questions the students must answer before embarking on the following project. These questions are designed to teach the student fundamentals of MATLAB and also the key concepts of DSP. For the latter part, each exercise includes a section summarizing the materials necessary for a quick review of DSP materials necessary to carry out the projects included in the exercise. For further details and explanations, each exercise includes at the end a list of DSP texts with specific chapter and/or section numbers. Each exercise also includes a section summarizing the MATLAB commands used to enable the student to find out more about one or more of these commands, if necessary, through the `help` command. A brief explanation of all MATLAB functions used in this book is given in Appendix B.

A novel feature of this book is the inclusion of partially written report documents for each of the first 10 laboratory exercises in the diskette provided. These reports are written in Microsoft Word. The students fill in the space provided answers to the questions as they proceed through the projects. This feature permits the students to complete more work in a specified amount of time than that would have been possible without it. The answers of the students should appear in a different font to make it easier for the laboratory instructor to evaluate the student's work. The completed reports also can serve as a guide for writing reports in other laboratory courses.

This book has evolved from teaching a laboratory component to an upper- division course on digital signal processing at the University of California, Santa Barbara, for the last 10 years. I would like to thank Drs. Stefan Thurnhofer and Ing-Song Lin for their assistance in developing the preliminary version of the laboratory course materials. I also would like to thank the students who took the upper division course and provided valuable comments that have improved the contents and style of the laboratory portion of the course. The complete manuscript of this book has been reviewed by Professor Hrvoje Babic of the University of Zagreb, Zagreb, Croatia; Professor Tamal Bose of the University of Colorado, Denver; Professor Ulrich Heute of the University of Kiel, Kiel, Germany; Professor Ottar Johnsen of the Ecole d'Ingénieurs de Friboug, Fribourg, Switzerland; Professor Abul N. Khondker of the Clarkson University, Potsdam, New York; Professor V. John Mathews of the Utah State University and Professor Yao Wang of the Polytechnic University, Brooklyn, New York. I thank them for their valuable comments. I thank my students, Rajeev Gandhi, Michael S. Moore and Debargha Mukherjee, for their assistance in proofreading of the manuscript and checking all the programs included in this book. I also acknowledge with gratitude the support of the Office of Instructional Development at the University of California, Santa Barbara, for providing me with two instructional improvement grants to develop the laboratory course. Finally, I thank my son Goutam for the cover design of my book.

Every attempt has been made to ensure the accuracy of all materials in this book, including the MATLAB programs. I would, however, appreciate readers bringing to my attention any errors that may have appeared in the printed version for reasons beyond my control and that of the publisher. These errors and any other comments can be communicated to me by e-mail addressed to: **mitra@ece.ucsb.edu**.

Santa Barbara   Sanjit K. Mitra

Discrete-Time Signals in the Time-Domain

1.1   Introduction

Digital signal processing is concerned with the processing of a discrete-time signal, called the *input signal*, to develop another discrete-time signal, called the *output signal*, with more desirable properties. In certain applications, it may be necessary to extract some key properties of the original signal using specific digital signal processing algorithms. It is also possible to investigate the properties of a discrete-time system by observing the output signals for specific input signals. It is thus important to learn first how to generate in the time-domain some basic discrete-time signals in MATLAB and perform elementary operations on them, which are the main objectives of this first exercise. A secondary objective is to learn the application of some basic MATLAB commands and how to apply them in simple digital signal processing problems.

1.2   Getting Started

The diskette provided with this book contains all of the MATLAB programs and the partially written reports for both the PC and the Macintosh PowerPC. In particular, it includes both PC and Macintosh versions of the MATLAB M-files of the first 10 exercises in folders grouped by chapters and report documents written in Microsoft Word in folders also grouped by chapters. After the completion of a project of a laboratory exercise, you record in the report of that exercise the answers to each question referring to this project at their designated locations.

Installation Instructions for a PC

To copy the program and the report folders onto the hard disk of a PC running Windows 95/98 follow the steps given below:

1. Insert the floppy disk.
2. Open the **My Computer** window by double-clicking on its icon displayed on the Desktop.
3. Open the window of the floppy drive by double-clicking on its icon.
4. Open the window of the desired hard drive by double-clicking on its icon. Depending on your setup, it may be necessary to open another window by double-

clicking on **My Computer** icon before you can select the destination hard drive.

5. In the floppy drive window, select the folder marked **PC** and drag it to the directory displayed in the hard drive window where you would like to copy the files.

Installation Instructions for a Macintosh PowerPC

To copy the program and the report folders on the hard disk of a PowerPC running Mac OS or higher follow the steps given below:

1. Insert the floppy disk.

2. Open the hard drive window by double-clicking on its icon displayed on the Desktop.

3. Open the window of the floppy drive by double-clicking on its icon.

4. In the floppy drive window, select the folder marked **MAC** and drag it to the directory displayed in the hard drive window where you would like to copy the files.

Downloading via FTP and the World Wide Web

The FTP site for downloading the files to a computer is **iplserv.ece.ucsb.edu**. The directories containing the files for the PC, Macintosh PowerPC, and UNIX workstations are as follows:

   pub/mitra/Labs/pc
   pub/mitra/Labs/mac
   pub/mitra/Labs/unix (M-files only)

To download the files from the FTP site to your PC or UNIX computer, follow the steps given below:

1. Open the FTP program for your computer.

2. Type **ftp iplserv.ece.ucsb.edu**.

3. At the **Login:** prompt, type **anonymous**.

4. At the **Password** prompt, type **Your E-mail address**.

5. To retrieve the PC files type **cd pub/mitra/Labs/pc**. To retrieve the UNIX files type **cd pub/mitra/Labs/unix**. These commands will get you to the directory where the M-files and the report files can be found.

6. Type **cd directory name** to change to the directory of the desired M-files or report files.

7. To download ASCII files such as the M-files, type **asc**. To download binary files such as the Word report documents (for the PC), type **bin**. To download a desired file type **get file name**. The last command will place the desired file in the current directory on your local system.

8. You can download another file using the **get** command or you can change back to the previous directory by typing **cd** followed by two periods.

9. When finished downloading all of the desired files, type **bye**.

To download the files from the FTP site to your Macintosh computer, a variety of FTP programs are available. The steps given below are for the **Fetch** program from Dartmouth College.

1. Open the FTP program.

2. Enter the following information in the **New Connection** dialog box. If you do not see this dialog box, open it by choosing **New Connection** from the **File** menu.

   > `Host:` **iplserv.ece.ucsb.edu**
   > `User ID:` **anonymous**
   > `Password:` **Your E-mail address**
   > `Directory` **/pub/mitra/Labs/mac**

3. Choose **OK**. You should see a window with the contents of the directory of the FTP site. You can change directories simply by double-clicking on the folder you want.

4. To transfer a file, select the file you want by double-clicking on it. A dialog box will pop up. Select the place on your computer where you want to store the file and choose **Save**. In recent versions of the **Fetch** program, entire directories can be downloaded at once by selecting the directory name and choosing **Get**.

For downloading via world wide web, follow the steps given below:

1. Open the available web browser.

2. Type **ftp://iplserv.ece.ucsb.edu** in the URL window.

3. Double-click on the desired directory (the directory for the PC and Macintosh versions are shown above).

4. Double-click on the desired file for downloading. You will get a dialog box asking where you would like to save the file.

1.3 Background Review

R1.1   A discrete-time signal is represented as a sequence of numbers, called *samples* . A sample value of a typical discrete-time signal or sequence $\{x[n]\}$ is denoted as $x[n]$ with the argument n being an integer in the range $-\infty$ and ∞. For convenience, the sequence $\{x[n]\}$ is often denoted without the curly brackets.

R1.2   The discrete-time signal may be a finite length or an infinite length sequence. A finite length (also called *finite duration* or *finite extent*) sequence is defined only for a finite time interval:

$$N_1 \leq n \leq N_2, \qquad (1.1)$$

where $-\infty < N_1$ and $N_2 < \infty$ with $N_2 \geq N_1$. The length or duration N of the finite length sequence is

$$N = N_2 - N_1 + 1. \qquad (1.2)$$

R1.3   A sequence $\tilde{x}[n]$ satisfying

$$\tilde{x}[n] = \tilde{x}[n + kN] \qquad \text{for all } n, \qquad (1.3)$$

is called a *periodic sequence* with a period N where N is a positive integer and k is any integer.

R1.4   The *energy* of a sequence $x[n]$ is defined by

$$\mathcal{E} = \sum_{n=-\infty}^{\infty} |x[n]|^2. \qquad (1.4)$$

The energy of a sequence over a finite interval $-K \leq n \leq K$ is defined by

$$\mathcal{E}_K = \sum_{n=-K}^{K} |x[n]|^2. \qquad (1.5)$$

R1.5   The *average power* of an aperiodic sequence $x[n]$ is defined by

$$\mathcal{P}_{av} = \lim_{K \to \infty} \frac{1}{2K+1} \mathcal{E}_K = \lim_{K \to \infty} \frac{1}{2K+1} \sum_{n=-K}^{K} |x[n]|^2. \qquad (1.6)$$

The average power of a periodic sequence $\tilde{x}[n]$ with a period N is given by

$$\mathcal{P}_{av} = \frac{1}{N} \sum_{n=0}^{N-1} |\tilde{x}[n]|^2. \qquad (1.7)$$

R1.6   The *unit sample sequence* , often called the *discrete-time impulse* or the *unit impulse*, denoted by $\delta[n]$, is defined by

$$\delta[n] = \begin{cases} 1, & \text{for } n = 0, \\ 0, & \text{for } n \neq 0. \end{cases} \qquad (1.8)$$

The *unit step sequence* , denoted by $\mu[n]$, is defined by

$$\mu[n] = \begin{cases} 1, & \text{for } n \geq 0, \\ 0, & \text{for } n < 0. \end{cases} \qquad (1.9)$$

R1.7 The *exponential sequence* is given by

$$x[n] = A\alpha^n, \tag{1.10}$$

where A and α are real or complex numbers. By expressing

$$\alpha = e^{(\sigma_o + j\omega_o)}, \quad \text{and} \quad A = |A|e^{j\phi},$$

we can rewrite Eq. (1.10) as

$$x[n] = |A|e^{\sigma_o n + j(\omega_o n + \phi)} = |A|e^{\sigma_o n}\cos(\omega_o n + \phi) + j|A|e^{\sigma_o n}\sin(\omega_o n + \phi). \tag{1.11}$$

R1.8 The *real sinusoidal sequence* with a constant amplitude is of the form

$$x[n] = A\cos(\omega_o n + \phi), \tag{1.12}$$

where A, ω_o, and ϕ are real numbers. The parameters A, ω_o, and ϕ in Eqs. (1.11) and (1.12) are called, respectively, the *amplitude* , the *angular frequency* , and the *initial phase* of the sinusoidal sequence $x[n]$. $f_o = \omega_o/2\pi$ is the *frequency*.

R1.9 The complex exponential sequence of Eq. (1.11) with $\sigma_o = 0$ and the sinusoidal sequence of Eq. (1.12) are periodic sequences if $\omega_o N$ is an integer multiple of 2π, that is,

$$\omega_o N = 2\pi r, \tag{1.13}$$

where N is a positive integer and r is any integer. The smallest possible N satisfying this condition is the *period* of the sequence.

R1.10 The *product* of two sequences $x[n]$ and $h[n]$ of length N yields a sequence $y[n]$, also of length N, as given by

$$y[n] = x[n] \cdot h[n]. \tag{1.14}$$

The *addition* of two sequences $x[n]$ and $h[n]$ of length N yields a sequence $y[n]$, also of length N, as given by

$$y[n] = x[n] + h[n]. \tag{1.15}$$

The *multiplication* of a sequence $x[n]$ of length N by a scalar A results in a sequence $y[n]$ of length N as given by

$$y[n] = A \cdot x[n]. \tag{1.16}$$

The *time-reversal* of a sequence $x[n]$ of infinite length results in a sequence $y[n]$ of infinite length as defined by

$$y[n] = x[-n]. \tag{1.17}$$

The *delay* of a sequence $x[n]$ of infinite length by a positive integer M results in a sequence $y[n]$ of infinite length given by

$$y[n] = x[n - M]. \tag{1.18}$$

If M is a negative integer, the operation indicated in Eq. (1.18) results in an *advance* of the sequence $x[n]$.

A sequence $x[n]$ of length N can be *appended* by another sequence $g[n]$ of length M resulting in a longer sequence $y[n]$ of length $N + M$ given by

$$\{y[n]\} = \{\{x[n]\}, \{g[n]\}\}. \tag{1.19}$$

1.4   MATLAB Commands Used

The MATLAB commands you will encounter in this exercise are as follows:

Operators and Special Characters

```
    :          .           +        –        *        /        ;
    %
```

Elementary Matrices and Matrix Manipulation

```
    i        ones        pi        rand        randn        zeros
```

Elementary Functions

```
    cos        exp        imag        real
```

Data Analysis

```
    sum
```

Two-Dimensional Graphics

```
    axis        grid        legend        plot        stairs
    stem        title       xlabel        ylabel
```

General Purpose Graphics Functions

```
    clf        subplot
```

Signal Processing Toolbox

```
    sawtooth        square
```

For additional information on these commands, see the *MATLAB Reference Guide* [Mat94] and the *Signal Processing Toolbox User's Guide* [Mat96] or type help commandname in the Command window. A brief explanation of the MATLAB functions used here can be found in Appendix B.

1.5   Generation of Sequences

The purpose of this section is to familiarize you with the basic commands in MATLAB for signal generation and for plotting the generated signal. MATLAB has been designed to operate on data stored as vectors or matrices. For our purposes, sequences will be stored as vectors. Therefore, all signals are limited to being causal and of finite length . The steps to follow to execute the programs listed in this book depend on the platform being used to run the MATLAB.

MATLAB on the Windows PC

The program can be executed by typing the name of the program without .m in the Command window and hitting the carriage return. Alternately, choose **Open** from the **File** menu in the Command window and choose the desired M-file. This opens the M-file in the Editor/Debugger window in which an M-file can be executed using the **Run** command under the **Tools** menu.

MATLAB on the Macintosh

The program can be executed by typing the name of the program without .m in the Command window and hitting the carriage return. Alternately, it can be copied into the Editor Window by using the **Open M-File** command on your screen and then choosing the **Save and Execute** command on your screen.

Project 1.1 Unit Sample and Unit Step Sequences

Two basic discrete-time sequences are the unit sample sequence and the unit step sequence of Eqs. (1.8) and (1.9), respectively. A unit sample sequence u[n] of length N can be generated using the MATLAB command

$$u = [1 \quad \text{zeros}(1, N-1)];$$

A unit sample sequence ud[n] of length N and delayed by M samples, where M < N, can be generated using the MATLAB command

$$ud = [\text{zeros}(1, M) \quad 1 \quad \text{zeros}(1, N-M-1)];$$

Likewise, a unit step sequence s[n] of length N can be generated using the MATLAB command

$$s = [\text{ones}(1, N)];$$

A delayed unit step sequence can be generated in a manner similar to that used in the generation of a delayed unit sample sequence.

Program P1_1 can be used to generate and plot a unit sample sequence.

```
% Program P1_1
% Generation of a Unit Sample Sequence
clf;
% Generate a vector from -10 to 20
n = -10:20;
% Generate the unit sample sequence
u = [zeros(1,10) 1 zeros(1,20)];
% Plot the unit sample sequence
stem(n,u);
xlabel('Time index n');ylabel('Amplitude');
title('Unit Sample Sequence');
axis([-10 20 0 1.2]);
```

Questions:

Q1.1   Run Program P1_1 to generate the unit sample sequence u[n] and display it.

Q1.2   What are the purposes of the commands clf, axis, title, xlabel, and ylabel?

Q1.3   Modify Program P1_1 to generate a delayed unit sample sequence ud[n] with a delay of 11 samples. Run the modified program and display the sequence generated.

Q1.4   Modify Program P1_1 to generate a unit step sequence s[n]. Run the modified program and display the sequence generated.

Q1.5   Modify Program P1_1 to generate a delayed unit step sequence sd[n] with an advance of 7 samples. Run the modified program and display the sequence generated.

Project 1.2   Exponential Signals

Another basic discrete-time sequence is the exponential sequence. Such a sequence can be generated using the MATLAB operators .^ and exp.

Program P1_2 given below can be employed to generate a complex-valued exponential sequence.

```
% Program P1_2
% Generation of a complex exponential sequence
clf;
c = -(1/12)+(pi/6)*i;
K = 2;
n = 0:40;
x = K*exp(c*n);
subplot(2,1,1);
stem(n,real(x));
xlabel('Time index n');ylabel('Amplitude');
title('Real part');
subplot(2,1,2);
stem(n,imag(x));
xlabel('Time index n');ylabel('Amplitude');
title('Imaginary part');
```

Program P1_3 given below can be employed to generate a real-valued exponential sequence.

```
% Program P1_3
% Generation of a real exponential sequence
clf;
n = 0:35; a = 1.2; K = 0.2;
x = K*a.^n;
```

```
stem(n,x);
xlabel('Time index n');ylabel('Amplitude');
```

Questions:

Q1.6 Run Program P1_2 and generate the complex-valued exponential sequence.

Q1.7 Which parameter controls the rate of growth or decay of this sequence? Which parameter controls the amplitude of this sequence?

Q1.8 What will happen if the parameter c is changed to (1/12)+(pi/6)*i?

Q1.9 What are the purposes of the operators real and imag?

Q1.10 What is the purpose of the command subplot?

Q1.11 Run Program P1_3 and generate the real-valued exponential sequence.

Q1.12 Which parameter controls the rate of growth or decay of this sequence? Which parameter controls the amplitude of this sequence?

Q1.13 What is the difference between the arithmetic operators ^ and .^?

Q1.14 What will happen if the parameter a is less than 1? Run Program P1_3 again with the parameter a changed to 0.9 and the parameter K changed to 20.

Q1.15 What is the length of this sequence and how can it be changed?

Q1.16 You can use the MATLAB command sum(s.*s) to compute the energy of a real sequence s[n] stored as a vector s. Evaluate the energy of the real-valued exponential sequences x[n] generated in Questions Q1.11 and Q1.14.

Project 1.3   Sinusoidal Sequences

Another very useful class of discrete-time signals is the real sinusoidal sequence of the form of Eq. (1.12). Such sinusoidal sequences can be generated in MATLAB using the trigonometric operators cos and sin.

Program P1_4 is a simple example that generates a sinusoidal signal.

```
% Program P1_4
% Generation of a sinusoidal sequence
n = 0:40;
f = 0.1;
phase = 0;
A = 1.5;
arg = 2*pi*f*n - phase;
x = A*cos(arg);
clf; % Clear old graph
stem(n,x); % Plot the generated sequence
```

```
axis([0 40 -2 2]);
grid;
title('Sinusoidal Sequence');
xlabel('Time index n');
ylabel('Amplitude');
axis;
```

Questions:

Q1.17   Run Program P1_4 to generate the sinusoidal sequence and display it.

Q1.18   What is the frequency of this sequence and how can it be changed? Which parameter controls the phase of this sequence? Which parameter controls the amplitude of this sequence? What is the period of this sequence?

Q1.19   What is the length of this sequence and how can it be changed?

Q1.20   Compute the average power of the generated sinusoidal sequence.

Q1.21   What are the purposes of the axis and grid commands?

Q1.22   Modify Program P1_4 to generate a sinusoidal sequence of frequency 0.9 and display it. Compare this new sequence with the one generated in Question Q1.17. Now, modify Program P1_4 to generate a sinusoidal sequence of frequency 1.1 and display it. Compare this new sequence with the one generated in Question Q1.17. Comment on your results.

Q1.23   Modify the above program to generate a sinusoidal sequence of length 50, frequency 0.08, amplitude 2.5, and phase shift 90 degrees and display it. What is the period of this sequence?

Q1.24   Replace the stem command in Program P1_4 with the plot command and run the program again. What is the difference between the new plot and the one generated in Question Q1.17?

Q1.25   Replace the stem command in Program P1_4 with the stairs command and run the program again. What is the difference between the new plot and those generated in Questions Q1.17 and Q1.24?

Project 1.4   Random Signals

A random signal of length N with samples uniformly distributed in the interval $(0,1)$ can be generated by using the MATLAB command

$$x = \text{rand}(1, N);$$

Likewise, a random signal x[n] of length N with samples normally distributed with zero mean and unity variance can be generated by using the following MATLAB command

$$x = \text{randn}(1, N);$$

Questions:

Q1.26 Write a MATLAB program to generate and display a random signal of length 100 whose elements are uniformly distributed in the interval $[-2, 2]$.

Q1.27 Write a MATLAB program to generate and display a Gaussian random signal of length 75 whose elements are normally distributed with zero mean and a variance of 3.

Q1.28 Write a MATLAB program to generate and display five sample sequences of a random sinusoidal signal of length 31

$$\{X[n]\} = \{A \cdot \cos(\omega_o n + \phi)\}, \tag{1.20}$$

where the amplitude A and the phase ϕ are statistically independent random variables with uniform probability distribution in the range $0 \leq A \leq 4$ for the amplitude and in the range $0 \leq \phi \leq 2\pi$ for the phase.

1.6 Simple Operations on Sequences

As indicated earlier, the purpose of digital signal processing is to generate a signal with more desirable properties from one or more given discrete-time signals. The processing algorithm consists of performing a combination of basic operations such as addition, scalar multiplication, time-reversal, delaying, and product operation (see R1.10). We consider here three very simple examples to illustrate the application of such operations .

Project 1.5   Signal Smoothing

A common example of a digital signal processing application is the removal of the noise component from a signal corrupted by additive noise. Let $s[n]$ be the signal corrupted by a random noise $d[n]$ resulting in the noisy signal $x[n] = s[n] + d[n]$. The objective is to operate on $x[n]$ to generate a signal $y[n]$ which is a reasonable approximation to $s[n]$. To this end, a simple approach is to generate an output sample by averaging a number of input samples around the sample at instant n. For example, a three-point moving average algorithm is given by

$$y[n] = \frac{1}{3}(x[n-1] + x[n] + x[n+1]). \tag{1.21}$$

Program P1_5 implements the above algorithm.

```
% Program P1_5
% Signal Smoothing by Averaging
clf;
R = 51;
d = 0.8*(rand(R,1) - 0.5); % Generate random noise
m = 0:R-1;
s = 2*m.*(0.9.^m); % Generate uncorrupted signal
x = s + d'; % Generate noise corrupted signal
```

```
subplot(2,1,1);
plot(m,d','r-',m,s,'g--',m,x,'b-.');
xlabel('Time index n');ylabel('Amplitude');
legend('d[n] ','s[n] ','x[n] ');
x1 = [0 0 x];x2 = [0 x 0];x3 = [x 0 0];
y = (x1 + x2 + x3)/3;
subplot(2,1,2);
plot(m,y(2:R+1),'r-',m,s,'g--');
legend('y[n] ','s[n] ');
xlabel('Time index n');ylabel('Amplitude');
```

Questions:

Q1.29   Run Program P1_5 and generate all pertinent signals.

Q1.30   What is the form of the uncorrupted signal s [n] ? What is the form of the additive noise d [n] ?

Q1.31   Can you use the statement x = s + d to generate the noise-corrupted signal? If not, why not?

Q1.32   What are the relations between the signals x1, x2, and x3, and the signal x?

Q1.33   What is the purpose of the legend command?

Project 1.6   Generation of Complex Signals

More complex signals can be generated by performing the basic operations on simple signals. For example, an *amplitude modulated signal* can be generated by modulating a high-frequency sinusoidal signal $x_H[n] = \cos(\omega_H n)$ with a low-frequency modulating signal $x_L[n] = \cos(\omega_L n)$. The resulting signal $y[n]$ is of the form

$$y[n] = A(1 + m \cdot x_L[n])x_H[n] = A(1 + m \cdot \cos(\omega_L n)) \cos(\omega_H n),$$

where m, called the *modulation index* , is a number chosen to ensure that $(1 + m \cdot x_L[n])$ is positive for all n. Program P1_6 can be used to generate an amplitude modulated signal.

```
% Program P1_6
% Generation of amplitude modulated sequence
clf;
n = 0:100;
m = 0.4;fH = 0.1; fL = 0.01;
xH = sin(2*pi*fH*n);
xL = sin(2*pi*fL*n);
y = (1+m*xL).*xH;
stem(n,y);grid;
xlabel('Time index n');ylabel('Amplitude');
```

Questions:

Q1.34 Run Program P1_6 and generate the amplitude modulated signal y[n] for various values of the frequencies of the carrier signal xH[n] and the modulating signal xL[n], and various values of the modulation index m.

Q1.35 What is the difference between the arithmetic operators * and .* ?

As the frequency of a sinusoidal signal is the derivative of its phase with respect to time, to generate a swept-frequency sinusoidal signal whose frequency increases linearly with time, the argument of the sinusoidal signal must be a quadratic function of time. Assume that the argument is of the form $an^2 + bn$ (i.e. the angular frequency is $2an + b$). Solve for the values of a and b from the given conditions (minimum angular frequency and maximum angular frequency). Program P1_7 is an example program to generate this kind of signal.

```
% Program P1_7
% Generation of a swept frequency sinusoidal sequence
n = 0:100;
a = pi/2/100;
b = 0;
arg = a*n.*n + b*n;
x = cos(arg);
clf;
stem(n, x);
axis([0,100,-1.5,1.5]);
title('Swept-Frequency Sinusoidal Signal');
xlabel('Time index n');
ylabel('Amplitude');
grid; axis;
```

Questions:

Q1.36 Run Program P1_7 and generate the swept-frequency sinusoidal sequence x[n].

Q1.37 What are the minimum and maximum frequencies of this signal?

Q1.38 How can you modify the above program to generate a swept sinusoidal signal with a minimum frequency of 0.1 and a maximum frequency of 0.3?

1.7   Workspace Information

The commands who and whos can be used to get information about the variables stored in the workspace and their sizes created in running various MATLAB programs at any time.

Questions:

Q1.39 Type who in the Command window. What information is displayed in the Command window as a result of this command?

Q1.40 Type whos in the Command window. What information is displayed in the Command window as a result of this command?

1.8   Other Types of Signals (Optional)

Project 1.7   Squarewave and Sawtooth Signals

MATLAB functions square and sawtooth can be used to generate sequences of the types shown in Figures 1.1 and 1.2, respectively.

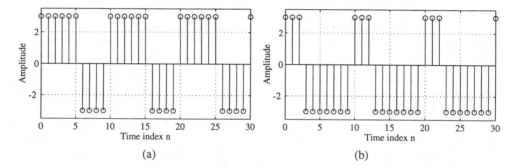

Figure 1.1   Square wave sequences.

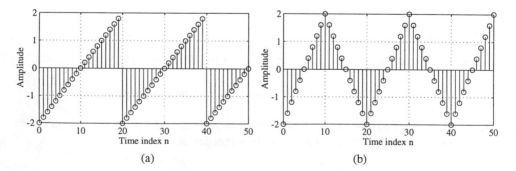

Figure 1.2   Sawtooth wave sequences.

Question:

Q1.41   Write MATLAB programs to generate the square-wave and the sawtooth wave sequences of the types shown in Figures 1.1 and 1.2. Using these programs, generate and plot the sequences.

1.9   Background Reading

[1] E. Cunningham. *Digital Filtering: An Introduction*. Houghton-Mifflin, Boston MA, 1992. Secs. 1.2–1.3.

[2] D. J. DeFatta, J. G. Lucas, and W. S. Hodgkiss. *Digital Signal Processing: A System Design Approach*. Wiley, New York NY, 1988. Secs. 2.1.2–2.1.4.

[3] L. B. Jackson. *Digital Filters and Signal Processing*. Kluwer, Boston MA, third edition, 1996. Secs. 2.2–2.3.

[4] R. Kuc. *Introduction to Digital Signal Processing*. McGraw-Hill, New York NY, 1988. Secs. 2-2, 2-4.

[5] L. C. Ludeman. *Fundamentals of Digital Signal Processing*. Harper & Row, New York NY, 1986. Sec. 1.2.

[6] S. K. Mitra. *Digital Signal Processing: A Computer-Based Approach*. McGraw-Hill, New York NY, 1998. Secs. 2.4–2.5.

[7] A. V. Oppenheim and R. W. Schafer. *Discrete-Time Signal Processing*. Prentice-Hall, Englewood Cliffs NJ, 1989. Sec. 2.1.

[8] B. Porat. *A Course in Digital Signal Procesing*. Wiley, New York NY, 1996. Sec. 2.7–2.8.

[9] J. G. Proakis and D. G. Manolakis. *Digital Signal Processing: Principles, Algorithms, and Applications*. Prentice-Hall, Englewood Cliffs NJ, 1996. Secs. 2.2–2.4.

[10] R. A. Roberts and C. T. Mullis. *Digital Signal Processing*, Addison-Wesley, Reading MA, 1987. Sec. 2.2.

Discrete-Time Systems in the Time-Domain

2.1 Introduction

A discrete-time system processes an input signal in the time-domain to generate an output signal with more desirable properties by applying an algorithm composed of simple operations on the input signal and its delayed versions. The aim of this second exercise is to illustrate the simulation of some simple discrete-time systems on the computer using MATLAB and investigate their time-domain properties.

2.2 Background Review

R2.1   For a *linear* discrete-time system , if $y_1[n]$ and $y_2[n]$ are the responses to the input sequences $x_1[n]$ and $x_2[n]$, respectively, then for an input

$$x[n] = \alpha\, x_1[n] + \beta\, x_2[n], \qquad (2.1)$$

the response is given by

$$y[n] = \alpha\, y_1[n] + \beta\, y_2[n]. \qquad (2.2)$$

The superposition property of Eq. (2.2) must hold for any arbitrary constants α and β and for all possible inputs $x_1[n]$ and $x_2[n]$. If Eq. (2.2) does not hold for at least one set of nonzero values of α and β, or one set of nonzero input sequences $x_1[n]$ and $x_2[n]$, then the system is *nonlinear* .

R2.2   For a *time-invariant* discrete-time system , if $y_1[n]$ is the response to an input $x_1[n]$, then the response to an input

$$x[n] = x_1[n - n_o]$$

is simply

$$y[n] = y_1[n - n_o].$$

where n_o is any positive or negative integer. The above relation between the input and output must hold for any arbitrary input sequence and its corresponding output. If it does not hold for at least one input sequence and its corresponding output sequence, the system is *time-varying* .

R2.3   A *linear time-invariant* (LTI) discrete-time system satisfies both the linearity and the time-invariance properties.

R2.4   If $y_1[n]$ and $y_2[n]$ are the responses of a *causal* discrete-time system to the inputs $u_1[n]$ and $u_2[n]$, respectively, then

$$u_1[n] = u_2[n] \quad \text{for } n < N$$

implies also that

$$y_1[n] = y_2[n] \quad \text{for } n < N.$$

R2.5   A discrete-time system is said to be *bounded-input, bounded-output* (BIBO) *stable* if, for any bounded input sequence $x[n]$, the corresponding output $y[n]$ is also a bounded sequence , that is, if

$$|x[n]| < B_x \quad \text{for all values of } n,$$

then the corresponding output $y[n]$ is also bounded, that is,

$$|y[n]| < B_y \quad \text{for all values of } n,$$

where B_x and B_y are finite constants.

R2.6   The response of a digital filter to a unit sample sequence $\{\delta[n]\}$ is called the *unit sample response* or, simply, the *impulse response* , and denoted as $\{h[n]\}$. Correspondingly, the response of a discrete-time system to a unit step sequence $\{\mu[n]\}$, denoted as $\{s[n]\}$, is its *unit step response* or, simply the *step response*.

R2.7   The response $y[n]$ of a linear, time-invariant discrete-time system characterized by an impulse response $h[n]$ to an input signal $x[n]$ is given by

$$y[n] = \sum_{k=-\infty}^{\infty} h[k]\,x[n-k], \tag{2.3}$$

which can be alternately written as

$$y[n] = \sum_{k=-\infty}^{\infty} h[n-k]\,x[k], \tag{2.4}$$

by a simple change of variables. The sum in Eqs. (2.3) and (2.4) is called the *convolution sum* of the sequences $x[n]$ and $h[n]$, and is represented compactly as:

$$y[n] = h[n] \circledast x[n], \tag{2.5}$$

where the notation $\circledast$ denotes the *convolution sum*.

R2.8   The overall impulse response $h[n]$ of the LTI discrete-time system obtained by a cascade connection of two LTI discrete-time systems with impulse responses $h_1[n]$ and $h_2[n]$, respectively, and as shown in Figure 2.1, is given by

$$h[n] = h_1[n] \circledast h_2[n]. \tag{2.6}$$

If the two LTI systems in the cascade connection of Figure 2.1 are such that

$$h_1[n] \circledast h_2[n] = \delta[n], \tag{2.7}$$

then the LTI system $h_2[n]$ is said to be the *inverse* of the LTI system $h_1[n]$ and vice-versa .

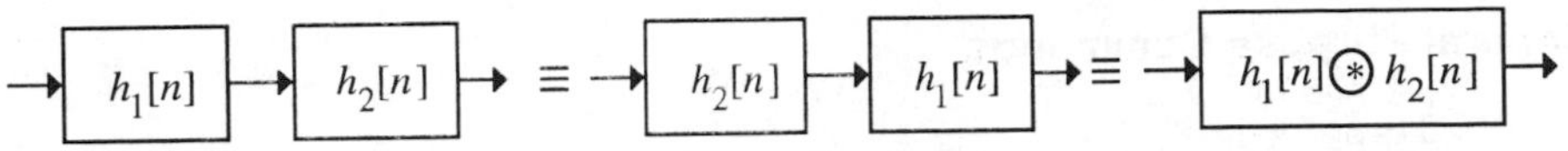

Figure 2.1 The cascade connection.

R2.9 An LTI discrete-time system is BIBO *stable* if and only if its impulse response sequence $\{h[n]\}$ is absolutely summable , that is,

$$\sum_{n=-\infty}^{\infty} |h[n]| < \infty. \tag{2.8}$$

R2.10 An LTI discrete-time system is *causal* if and only if its impulse response sequence $\{h[n]\}$ satisfies the condition

$$h[k] = 0 \quad \text{for} \quad k < 0. \tag{2.9}$$

R2.11 The class of LTI discrete-time systems with which we shall be mostly concerned in this book is characterized by a linear constant coefficient difference equation of the form

$$\sum_{k=0}^{N} d_k \, y[n-k] = \sum_{k=0}^{M} p_k \, x[n-k], \tag{2.10}$$

where $x[n]$ and $y[n]$ are, respectively, the input and the output of the system, and $\{d_k\}$ and $\{p_k\}$ are constants. The *order* of the discrete-time system is $\max(N, M)$, which is the order of the difference equation characterizing the system. If we assume the system to be causal, then we can rewrite Eq. (2.9) to express $y[n]$ explicitly as a function of $x[n]$:

$$y[n] = -\sum_{k=1}^{N} \frac{d_k}{d_0} y[n-k] + \sum_{k=0}^{M} \frac{p_k}{d_0} x[n-k], \tag{2.11}$$

provided $d_0 \neq 0$. The output $y[n]$ can be computed using Eq. (2.10) for all $n \geq n_o$ knowing $x[n]$ and the *initial conditions* $y[n_o - 1], y[n_o - 2], \ldots, y[n_o - N]$.

R2.12 A discrete-time system is called a *finite impulse response* (FIR) system if its impulse response $h[n]$ is of finite length. Otherwise, it is an *infinite impulse response* (IIR) system . The causal system of Eq. (2.11) represents an FIR system if $d_k = 0$ for $k > 0$. Otherwise, it is an IIR system.

2.3   MATLAB Commands Used

The MATLAB commands you will encounter in this exercise are as follows:

General Purpose Commands

```
disp
```

Operators and Special Characters

```
:          .          +          –          *          /          ;
%          <
```

Language Constructs and Debugging

```
break       end        for        if        input
```

Elementary Matrices and Matrix Manipulation

```
ones        pi         zeros
```

Elementary Functions

```
abs         cos
```

Polynomial and Interpolation Functions

```
conv
```

Two-Dimensional Graphics

```
axis        plot       stem       title       xlabel
ylabel
```

General Purpose Graphics Functions

```
clf         subplot
```

Character String Functions

```
num2str
```

Signal Processing Toolbox

```
filter      impz
```

For additional information on these commands, see the *MATLAB Reference Guide* [Mat94] and the *Signal Processing Toolbox User's Guide* [Mat96] or type help commandname in the Command window. A brief explanation of the MATLAB functions used here can be found in Appendix B.

2.4   Simulation of Discrete-Time Systems

In Project 1.5 we illustrated the application of a simple discrete-time system described by Eq. (1.21) in the smoothing of data corrupted by a random noise . We now consider the simulation of some additional discrete-time systems and study their properties. For the simulation of causal LTI discrete-time systems described by Eq. (2.10), the command `filter` can be used. There are several versions of this command. If we denote

$$\text{num} = [p_0 \quad p_1 \quad \cdots \quad p_M],$$

$$\text{den} = [d_0 \quad d_1 \quad \cdots \quad d_N],$$

then `y = filter(num,den,x)` generates an output vector y of the same length as the specified input vector x with zero initial conditions, that is, `y[-1] = y[-2] = ... = y[-N] = 0`. The output can also be computed using `y = filter(num,den,x, ic)` where `ic = [y[-1], y[-2], ..., y[-N]]` is the vector of initial conditions. Access to final conditions is obtained using `[y, fc] = filter(num,den,x, ic)`.

Project 2.1   The Moving Average System

Examination of Eq. (1.21) reveals that the three-point smoothing filter considered here is an LTI FIR system. Moreover, as $y[n]$ depends on a future input sample $x[n + 1]$, the system is noncausal. A causal version of the three-point smoothing filter is obtained by simply delaying the output by one sample period, resulting in the FIR filter described by

$$y[n] = \frac{1}{3}(x[n] + x[n - 1] + x[n - 2]). \tag{2.12}$$

Generalizing the above equation we obtain

$$y[n] = \frac{1}{M} \sum_{k=0}^{M-1} x[n - k], \tag{2.13}$$

which defines a causal M-point smoothing FIR filter. The system of Eq. (2.13) is also known as a *moving average filter* . We illustrate its use in filtering high-frequency components from a signal composed of a sum of several sinusoidal signals.

```
% Program P2_1
% Simulation of an M-point Moving Average Filter
% Generate the input signal
clf;
n = 0:100;
s1 = cos(2*pi*0.05*n);% A low-frequency sinusoid
s2 = cos(2*pi*0.47*n); % A high frequency sinusoid
x = s1+s2;
% Implementation of the moving average filter
M = input('Desired length of the filter = ');
num = ones(1,M);
```

```
y = filter(num,1,x)/M;
% Display the input and output signals
subplot(2,2,1);
plot(n, s1);
axis([0, 100, -2, 2]);
xlabel('Time index n'); ylabel('Amplitude');
title('Signal # 1');
subplot(2,2,2);
plot(n, s2);
axis([0, 100, -2, 2]);
xlabel('Time index n'); ylabel('Amplitude');
title('Signal # 2');
subplot(2,2,3);
plot(n, x);
axis([0, 100, -2, 2]);
xlabel('Time index n'); ylabel('Amplitude');
title('Input Signal');
subplot(2,2,4);
plot(n, y);
axis([0, 100, -2, 2]);
xlabel('Time index n'); ylabel('Amplitude');
title('Output Signal');
axis;
```

Questions:

Q2.1   Run the above program for M = 2 to generate the output signal with x[n] = s1[n] + s2[n] as the input. Which component of the input x[n] is suppressed by the discrete-time system simulated by this program?

Q2.2   If the LTI system is changed from y[n] = 0.5(x[n] + x[n - 1]) to y[n] = 0.5(x[n] - x[n - 1]), what would be its effect on the input x[n] = s1[n] + s2[n]?

Q2.3   Run Program P2_1 for other values of filter length M, and various values of the frequencies of the sinusoidal signals s1[n] and s2[n]. Comment on your results.

Q2.4   Modify Program P2_1 to use a swept-frequency sinusoidal signal of length 101, a minimum frequency 0, and a maximum frequency 0.5 as the input signal (see Program P1_7) and compute the output signal. Can you explain the results of Questions Q2.1 and Q2.2 from the response of this system to the swept-frequency signal ?

Project 2.2   A Simple Nonlinear Discrete-Time System (Optional)

Let $y[n]$ be a signal generated by applying the following nonlinear operations on a signal $x[n]$:

$$y[n] = x[n]^2 - x[n - 1]\, x[n + 1]. \tag{2.14}$$

In this project you will generate the output $y[n]$ of the above system for different types of the input $x[n]$ using Program P2_2.

The following MATLAB program can be used to generate an input signal $x[n]$ composed of a sum of two sinusoidal sequences and simulate the LTI system of Eq. (2.12) to generate $y[n]$.

```
% Program P2_2
% Generate a sinusoidal input signal
clf;
n = 0:200;
x = cos(2*pi*0.05*n);
% Compute the output signal
x1 = [x 0 0]; % x1[n] = x[n+1]
x2 = [0 x 0]; % x2[n] = x[n]
x3 = [0 0 x]; % x3[n] = x[n-1]
y = x2.*x2 - x1.*x3;
y = y(2:202);
% Plot the input and output signals
subplot(2,1,1)
plot(n, x)
xlabel('Time index n');ylabel('Amplitude');
title('Input Signal')
subplot(2,1,2)
plot(n,y)
xlabel('Time index n');ylabel('Amplitude');
title('Output signal');
```

Questions:

Q2.5 Use sinusoidal signals with different frequencies as the input signals and compute the output signal for each input. How do the output signals depend on the frequencies of the input signal? Can you verify your observation mathematically?

Q2.6 Use sinusoidal signals of the form $x[n] = \sin(\omega_o n) + K$ as the input signal and compute the output signal. How does the output signal $y[n]$ depend on the DC value K?

Project 2.3   Linear and Nonlinear Systems

We now investigate the linearity property (see R2.1) of a causal system of the type described by Eq. (2.10) . Consider the system given by

$$y[n]-0.4\,y[n-1]+0.75\,y[n-2] = 2.2403\,x[n]+2.4908\,x[n-1]+2.2403\,x[n-2]. \quad (2.15)$$

MATLAB Program P2_3 is used to simulate the system of Eq. (2.15), to generate three different input sequences $x_1[n]$, $x_2[n]$, and $x[n] = a \cdot x_1[n] + b \cdot x_2[n]$, and to compute and plot the corresponding output sequences $y_1[n]$, $y_2[n]$, and $y[n]$.

```
% Program P2_3
% Generate the input sequences
clf;
n = 0:40;
a = 2;b = -3;
x1 = cos(2*pi*0.1*n);
x2 = cos(2*pi*0.4*n);
x = a*x1 + b*x2;
num = [2.2403 2.4908 2.2403];
den = [1 -0.4 0.75];
ic = [0 0]; % Set zero initial conditions
y1 = filter(num,den,x1,ic); % Compute the output y1[n]
y2 = filter(num,den,x2,ic); % Compute the output y2[n]
y = filter(num,den,x,ic); % Compute the output y[n]
yt = a*y1 + b*y2;
d = y - yt; % Compute the difference output d[n]
% Plot the outputs and the difference signal
subplot(3,1,1)
stem(n,y);
ylabel('Amplitude');
title('Output Due to Weighted Input:  a \cdot x_{1}[n]
+ b \cdot x_{2}[n]');
subplot(3,1,2)
stem(n,yt);
ylabel('Amplitude');
title('Weighted Output:  a \cdot y_{1}[n] + b \cdot y_{2}[n]');
subplot(3,1,3)
stem(n,d);
xlabel('Time index n'); ylabel('Amplitude');
title('Difference Signal');
```

Questions:

Q2.7 Run Program P2_3 and compare y[n] obtained with weighted input with yt[n] obtained by combining the two outputs y1[n] and y2[n] with the same weights. Are these two sequences equal? Is this system linear?

Q2.8 Repeat Question Q2.7 for three different sets of values of the weighting constants, a and b, and three different sets of input frequencies.

Q2.9 Repeat Question Q2.7 with nonzero initial conditions .

Q2.10 Repeat Question Q2.8 with nonzero initial conditions.

Q2.11 Consider another system described by:

$$y[n] = x[n]\,x[n-1].$$

Modify Program P2_3 to compute the output sequences y1[n], y2[n], and y[n] of the above system. Compare y[n] with yt[n]. Are these two sequences equal? Is this system linear?

Project 2.4 Time-Invariant and Time-Varying Systems

We next investigate the time-invariance property (see R2.2) of a causal system of the type described by Eq. (2.11). Consider again the system given by Eq. (2.15).

MATLAB Program P2_4 is used to simulate the system of Eq. (2.15), to generate two different input sequences x[n] and x[n - D], and to compute and plot the corresponding output sequences y1[n], y2[n], and the difference y1[n] - y2[n + D].

```
% Program P2_4
% Generate the input sequences
clf;
n = 0:40; D = 10;a = 3.0;b = -2;
x = a*cos(2*pi*0.1*n) + b*cos(2*pi*0.4*n);
xd = [zeros(1,D) x];
num = [2.2403 2.4908 2.2403];
den = [1 -0.4 0.75];
ic = [0 0];% Set initial conditions
% Compute the output y[n]
y = filter(num,den,x,ic);
% Compute the output yd[n]
yd = filter(num,den,xd,ic);
% Compute the difference output d[n]
d = y - yd(1+D:41+D);
% Plot the outputs
subplot(3,1,1)
stem(n,y);
ylabel('Amplitude');
title('Output y[n]');grid;
subplot(3,1,2)
stem(n,yd(1:41));
ylabel('Amplitude');
title(['Output Due to Delayed Input x[n ', num2str(D),']']);grid;
subplot(3,1,3)
stem(n,d);
xlabel('Time index n'); ylabel('Amplitude');
title('Difference Signal');grid;
```

Questions:

Q2.12   Run Program P2_4 and compare the output sequences y[n] and yd[n - 10]. What is the relation between these two sequences? Is this system time-invariant?

Q2.13   Repeat Question Q2.12 for three different values of the delay variable D.

Q2.14   Repeat Question Q2.12 for three different sets of values of the input frequencies.

Q2.15   Repeat Question Q2.12 for nonzero initial conditions . Is this system time-invariant?

Q2.16   Repeat Question Q2.14 for nonzero initial conditions. Is this system time-invariant?

Q2.17   Consider another system described by:

$$y[n] = n\,x[n] + x[n-1]. \qquad (2.16)$$

Modify Program P2_4 to simulate the above system and determine whether this system is time-invariant or not.

Q2.18   (optional) Modify Program P2_3 to test the linearity of the system of Eq. (2.16).

2.5   Linear Time-Invariant Discrete-Time Systems

Project 2.5   Computation of Impulse Responses of LTI Systems

The MATLAB command y = impz(num,den,N) can be used to compute the first N samples of the impulse response of the causal LTI discrete-time system of Eq. (2.11).

MATLAB Program P2_5 given below computes and plots the impulse response of the system described by Eq. (2.15).

```
% Program P2_5
% Compute the impulse response y
clf;
N = 40;
num = [2.2403 2.4908 2.2403];
den = [1 -0.4 0.75];
y = impz(num,den,N);
% Plot the impulse response
stem(y);
xlabel('Time index n'); ylabel('Amplitude');
title('Impulse Response'); grid;
```

Questions:

Q2.19   Run Program P2_5 and generate the impulse response of the discrete-time system of Eq. (2.15).

Q2.20   Modify Program P2_5 to generate the first 45 samples of the impulse response of the following causal LTI system:

$$y[n] + 0.71\,y[n-1] - 0.46\,y[n-2] - 0.62\,y[n-3]$$
$$= 0.9\,x[n] - 0.45\,x[n-1] + 0.35\,x[n-2] + 0.002\,x[n-3]. \qquad (2.17)$$

Q2.21 Write a MATLAB program to generate the impulse response of a causal LTI system of Eq. (2.17) using the `filter` command; compute and plot the first 40 samples. Compare your result with that obtained in Question Q2.20.

Q2.22 Write a MATLAB program to generate and plot the step response of a causal LTI system of Eq. (2.11). Using this program compute and plot the first 40 samples of the step response of the LTI system of Eq. (2.15).

Project 2.6   Cascade of LTI Systems

In practice a causal LTI discrete-time system of higher order is implemented as a cascade of lower order causal LTI discrete-time systems. For example, the fourth-order discrete-time system given below

$$y[n] + 1.6\,y[n-1] + 2.28\,y[n-2] + 1.325\,y[n-3] + 0.68\,y[n-4]$$
$$= 0.06\,x[n] - 0.19\,x[n-1] + 0.27\,x[n-2] - 0.26\,x[n-3] + 0.12\,x[n-4] \quad (2.18)$$

can be realized as a cascade of two second-order discrete-time systems:

Stage No. 1

$$y_1[n] + 0.9\,y_1[n-1] + 0.8\,y_1[n-2] = 0.3\,x[n] - 0.3\,x[n-1] + 0.4\,x[n-2] \quad (2.19)$$

Stage No. 2

$$y_2[n] + 0.7\,y_2[n-1] + 0.85\,y_2[n-2] = 0.2\,y_1[n] - 0.5\,y_1[n-1] + 0.3\,y_1[n-2] \quad (2.20)$$

MATLAB Program P2.6 simulates the fourth-order system of Eq. (2.18), and the cascade system of Eqs. (2.19) and (2.20). It first generates a sequence x[n], and then uses it as the input of the fourth-order system, generating the output y[n]. It then applies the same input x[n] to Stage No. 1 and finds its output sequence y1[n]. Next, it uses y1[n] as the input of Stage No. 2 and finds its output y2[n]. Finally, the difference between the two overall outputs y[n] and y2[n] are formed. All output and the difference signals are then plotted.

```
% Program P2_6
% Cascade Realization
clf;
x = [1 zeros(1,40)];% Generate the input
n = 0:40;
% Coefficients of 4th-order system
den = [1 1.6 2.28 1.325 0.68];
num = [0.06 -0.19 0.27 -0.26 0.12];
% Compute the output of 4th-order system
y = filter(num,den,x);
% Coefficients of the two 2nd-order systems
num1 = [0.3 -0.2 0.4];den1 = [1 0.9 0.8];
```

```
num2 = [0.2 -0.5 0.3];den2 = [1 0.7 0.85];
% Output y1[n] of the first stage in the cascade
y1 = filter(num1,den1,x);
% Output y2[n] of the second stage in the cascade
y2 = filter(num2,den2,y1);
% Difference between y[n] and y2[n]
d = y - y2;
% Plot output and difference signals
subplot(3,1,1);
stem(n,y);
ylabel('Amplitude');
title('Output of 4th-order Realization');grid;
subplot(3,1,2);
stem(n,y2)
ylabel('Amplitude');
title('Output of Cascade Realization');grid;
subplot(3,1,3);
stem(n,d)
xlabel('Time index n');ylabel('Amplitude');
title('Difference Signal');grid;
```

Questions:

Q2.23   Run Program P2_6 to compute the output sequences y[n] and y2[n] and the difference signal d[n]. Is y[n] the same as y2[n]?

Q2.24   Repeat Question Q2.23 with the input changed to a sinusoidal sequence.

Q2.25   Repeat Question Q2.23 with arbitrary nonzero initial condition vectors ic, ic1, and ic2.

Q2.26   Modify Program P2_6 to repeat the same procedure with the two second-order systems in reverse order and with zero initial conditions . Is there any difference between the two outputs?

Q2.27   Repeat Question Q2.26 with arbitrary nonzero initial condition vectors ic, ic1, and ic2.

Project 2.7   Convolution

The convolution operation of Eq. (2.5) is implemented in MATLAB by the command conv, provided the two sequences to be convolved are of finite length. For example, the output sequence of an FIR system can be computed by convolving its impulse response with a given finite-length input sequence. The following MATLAB program illustrates this approach.

```
% Program P2_7
clf;
```

```
h = [3 2 1 -2 1 0 -4 0 3]; % impulse response
x = [1 -2 3 -4 3 2 1]; % input sequence
y = conv(h,x);
n = 0:14;
subplot(2,1,1);
stem(n,y);
xlabel('Time index n'); ylabel('Amplitude');
title('Output Obtained by Convolution');grid;
x1 = [x zeros(1,8)];
y1 = filter(h,1,x1);
subplot(2,1,2);
stem(n,y1);
xlabel('Time index n'); ylabel('Amplitude');
title('Output Generated by Filtering');grid;
```

Questions:

Q2.28 Run Program P2_7 to generate y[n] obtained by the convolution of the sequences h[n] and x[n], and to generate y1[n] obtained by filtering the input x[n] by the FIR filter h[n]. Is there any difference between y[n] and y1[n]? What is the reason for using x1[n] obtained by zero-padding x[n] as the input for generating y1[n]?

Q2.29 Modify Program P2_7 to develop the convolution of a length-15 sequence h[n] with a length-10 sequence x[n], and repeat Question Q2.28. Use your own sample values for h[n] and x[n].

Project 2.8   Stability of LTI Systems

As indicated by Eq. (2.8), an LTI discrete-time system is BIBO stable if its impulse response is absolutely summable. It therefore follows that a necessary condition for an IIR LTI system to be stable is that its impulse response decays to zero as the sample index gets larger. Program P2_8 is a simple MATLAB program used to compute the sum of the absolute values of the impulse response samples of a causal IIR LTI system. It computes N samples of the impulse response sequence, evaluates

$$S(K) = \sum_{n=0}^{K} |h[n]| \tag{2.21}$$

for increasing values of K, and checks the value of $|h(K)|$ at each iteration step. If the value of $|h[K]|$ is smaller than 10^{-6}, then it is assumed that the sum $S(K)$ of Eq. (2.21) has converged and is very close to $S(\infty)$.

```
% Program P2_8
% Stability test based on the sum of the absolute
% values of the impulse response samples
clf;
```

```
num = [1 -0.8]; den = [1 1.5 0.9];
N = 200;
h = impz(num,den,N+1);
parsum = 0;
for k = 1:N+1;
    parsum = parsum + abs(h(k));
    if abs(h(k)) < 10^(-6), break, end
end
% Plot the impulse response
n = 0:N;
stem(n,h)
xlabel('Time index n'); ylabel('Amplitude');
disp('Value =');disp(abs(h(k))); % Print the value of abs(h(k))
```

Questions:

Q2.30   What are the purposes of the commands `for` and `end`?

Q2.31   What is the purpose of the command `break`?

Q2.32   What is the discrete-time system whose impulse response is being determined by Program P2_8? Run Program P2_8 to generate the impulse response. Is this system stable? If $|h[K]|$ is not smaller than 10^{-6} but the plot shows a decaying impulse response, run Program P2_8 again with a larger value of N.

Q2.33   Consider the following discrete-time system characterized by the difference equation:

$$y[n] = x[n] - 4\,x[n-1] + 3\,x[n-2] + 1.7\,y[n-1] - y[n-2].$$

Modify Program P2_8 to compute and plot the impulse response of the above system. Is this system stable?

Project 2.9   Illustration of the Filtering Concept

Consider the following two discrete-time systems characterized by the difference equations:

System No. 1

$$y[n] = 0.5\,x[n] + 0.27\,x[n-1] + 0.77\,x[n-2],$$

System No. 2

$$y[n] = 0.45\,x[n] + 0.5\,x[n-1] + 0.45\,x[n-2] + 0.53\,y[n-1] - 0.46\,y[n-2].$$

MATLAB Program P2_9 is used to compute the outputs of the above two systems for an input

$$x[n] = \cos\left(\frac{20\pi n}{256}\right) + \cos\left(\frac{200\pi n}{256}\right), \quad \text{with} \quad 0 \le n < 299.$$

```
% Program P2_9
% Generate the input sequence
clf;
n = 0:299;
x1 = cos(2*pi*10*n/256);
x2 = cos(2*pi*100*n/256);
x = x1+x2;
% Compute the output sequences
num1 = [0.5 0.27 0.77];
y1 = filter(num1,1,x); % Output of System No. 1
den2 = [1 -0.53 0.46];
num2 = [0.45 0.5 0.45];
y2 = filter(num2,den2,x); % Output of System No.
% Plot the output sequences
subplot(2,1,1);
plot(n,y1);axis([0 300 -2 2]);
ylabel('Amplitude');
title('Output of System No. 1');grid;
subplot(2,1,2);
plot(n,y2);axis([0 300 -2 2]);
xlabel('Time index n'); ylabel('Amplitude');
title('Output of System No. 2');grid;
```

Questions:

Q2.34 Run Program P2_9. In this question both filters are lowpass filters but with different attenuation in the stopband, especially at the frequencies of the input signal. Which filter has better characteristics for suppression of the high-frequency component of the input signal $x[n]$?

Q2.35 Modify Program P2_9 by changing the input sequence to a swept sinusoidal sequence (length 301, minimum frequency 0, and maximum frequency 0.5). Which filter has better characteristics for suppression of the high-frequency component?

2.6   Background Reading

[1] A. Antoniou. *Digital Filters: Analysis, Design, and Applications*. McGraw-Hill, New York NY, second edition, 1993. Secs. 1.3 – 1.8.

[2] E. Cunningham. *Digital Filtering: An Introduction*. Houghton-Mifflin, New York NY, 1992. Ch. 8.

[3] D.J. DeFatta, J.G. Lucas, and W.S. Hodgkiss. *Digital Signal Processing: A System Design Approach*. Wiley, New York NY, 1988. Sec. 2.2.

[4] L.B. Jackson. *Digital Filters and Signal Processing.* Kluwer, Boston MA, third edition, 1996. Ch. 11.

[5] R. Kuc. *Introduction to Digital Signal Processing.* McGraw-Hill, New York NY, 1988. Secs. 2-3, 2-5 – 2-8.

[6] L.C. Ludeman. *Fundamentals of Digital Signal Processing,* Harper & Row, New York NY, 1986. Sec. 1.3.

[7] S.K. Mitra. *Digital Signal Processing: A Computer-Based Approach.* McGraw-Hill, New York NY, 1998. Sec. 8.4 and Ch. 9.

[8] A.V. Oppenheim and R.W. Schafer. *Discrete-Time Signal Processing.* Prentice-Hall, Englewood Cliffs NJ, 1989. Secs. 2.2–2.5.

[9] S.J. Orfanidis. *Introduction to Signal Processing.* Prentice-Hall, Englewood Cliffs NJ, 1996. Chs. 2, 3.

[10] B. Porat. *A Course in Digital Signal Procesing.* Wiley, New York, NY, 1996. Sec. 2.7.

[11] J.G. Proakis and D.G. Manolakis. *Digital Signal Processing: Principles, Algorithms, and Applications.* Prentice-Hall, Englewood Cliffs NJ, third edition, 1996. Secs. 2.2–2.4.

[12] R.A. Roberts and C.T. Mullis. *Digital Signal Processing.* Addison-Wesley, Reading MA, 1987. Secs. 2.3–2.4.

Discrete-Time Signals in the Frequency-Domain

3

3.1   Introduction

In the previous two exercises you dealt with the time-domain representation of discrete-time signals and systems, and investigated their properties. Further insight into the properties of such signals and systems is obtained by their representation in the frequency-domain. To this end three commonly used representations are the discrete-time Fourier transform (DTFT), the discrete Fourier transform (DFT), and the z-transform. In this exercise you will study all three representations of a discrete-time sequence.

3.2   Background Review

R3.1   The *discrete-time Fourier transform* (DTFT) $X(e^{j\omega})$ of a sequence $x[n]$ is defined by

$$X(e^{j\omega}) = \sum_{n=-\infty}^{\infty} x[n]e^{-j\omega n}. \qquad (3.1)$$

In general $X(e^{j\omega})$ is a complex function of the real variable ω and can be written as

$$X(e^{j\omega}) = X_{re}(e^{j\omega}) + jX_{im}(e^{j\omega}), \qquad (3.2)$$

where $X_{re}(e^{j\omega})$ and $X_{im}(e^{j\omega})$ are, respectively, the real and imaginary parts of $X(e^{j\omega})$, and are real functions of ω. $X(e^{j\omega})$ can alternately be expressed in the form

$$X(e^{j\omega}) = |X(e^{j\omega})|e^{j\theta(\omega)}, \qquad (3.3)$$

where

$$\theta(\omega) = arg\{X(e^{j\omega})\}. \qquad (3.4)$$

The quantity $|X(e^{j\omega})|$ is called the *magnitude function* and the quantity $\theta(\omega)$ is called the *phase function* , with both functions again being real functions of ω. In many applications, the Fourier transform is called the *Fourier spectrum* and, likewise, $|X(e^{j\omega})|$ and $\theta(\omega)$ are referred to as the *magnitude spectrum* and *phase spectrum*, respectively.

R3.2   The DTFT $X(e^{j\omega})$ is a periodic continuous function in ω with a period 2π.

R3.3   For a real sequence $x[n]$, the real part $X_{re}(e^{j\omega})$ of its DTFT and the magnitude function $|X(e^{j\omega})|$ are even functions of ω, whereas the imaginary part $X_{im}(e^{j\omega})$ and the phase function $\theta(\omega)$ are odd functions of ω.

33

R3.4   The *inverse discrete-time Fourier transform* $x[n]$ of $X(e^{j\omega})$ is given by

$$x[n] = \frac{1}{2\pi} \int_{-\pi}^{\pi} X(e^{j\omega}) d\omega. \tag{3.5}$$

R3.5   The Fourier transform $X(e^{j\omega})$ of a sequence $x[n]$ exists if $x[n]$ is *absolutely summable*, that is,

$$\sum_{n=-\infty}^{\infty} |x[n]| < \infty, \tag{3.6}$$

R3.6   The DTFT satisfies a number of useful properties that are often uitilized in a number of applications. A detailed listing of these properties and their analytical proofs can be found in any text on digital signal processing. These properties can also be verified using MATLAB. We list below a few selected properties that will be encountered later in this exercise.

Time-Shifting Property – If $G(e^{j\omega})$ denotes the DTFT of a sequence $g[n]$, then the DTFT of the time-shifted sequence $g[n - n_o]$ is given by $e^{-j\omega n_o} G(e^{j\omega})$.

Frequency-Shifting Property – If $G(e^{j\omega})$ denotes the DTFT of a sequence $g[n]$, then the DTFT of the sequence $e^{j\omega_o n} g[n]$ is given by $G(e^{j(\omega - \omega_o)})$.

Convolution Property – If $G(e^{j\omega})$ and $H(e^{j\omega})$ denote the DTFTs of the sequences $g[n]$ and $h[n]$, respectively, then the DTFT of the sequence $g[n] \star h[n]$ is given by $G(e^{j\omega})H(e^{j\omega})$.

Modulation Property – If $G(e^{j\omega})$ and $H(e^{j\omega})$ denote the DTFTs of the sequences $g[n]$ and $h[n]$, respectively, then the DTFT of the sequence $g[n]h[n]$ is given by

$$\frac{1}{2\pi} \int_{-\pi}^{\pi} G(e^{j\theta}) H(e^{j(\omega - \theta)}) d\theta.$$

Time-Reversal Property – If $G(e^{j\omega})$ denotes the DTFT of a sequence $g[n]$, then the DTFT of the time-reversed sequence $g[-n]$ is given by $G(e^{-j\omega})$.

R3.7   The N-point *discrete Fourier transform* (DFT) of a finite-length sequence $x[n]$, defined for $0 \leq n \leq N - 1$, is given by

$$X[k] = \sum_{n=0}^{N-1} x[n] W_N^{kn}, \quad k = 0, 1, \ldots, N - 1, \tag{3.7}$$

where

$$W_N = e^{-j2\pi/N}. \tag{3.8}$$

R3.8   The N-point DFT $X[k]$ of a length-N sequence $x[n], n = 0, 1, \ldots, N-1$, is simply the frequency samples of its DTFT $X(e^{j\omega})$ evaluated at N uniformly spaced frequency points, $\omega = \omega_k = 2\pi k/N, k = 0, 1, \ldots, N - 1$, that is,

$$X[k] = X(e^{j\omega})|_{\omega=2\pi k/N}, \quad k = 0, 1, \ldots, N - 1. \tag{3.9}$$

R3.9   The N-point *circular convolution* of two length-N sequences $g[n]$ and $h[n]$, $0 \le n \le N - 1$, is defined by

$$y_C[n] = \sum_{m=0}^{N-1} g[m]h[\langle n - m \rangle_N], \qquad (3.10)$$

where $\langle n \rangle_N = n$ modulo N. The N-point circular convolution operation is usually denoted as

$$y_C[n] = g[n] \, \circledN \, h[n]. \qquad (3.11)$$

R3.10   The linear convolution of a length-N sequence $g[n]$, $0 \le n \le N - 1$, with a length$-M$ sequence $h[n]$, $0 \le n \le M - 1$, can be obtained by a $(N + M - 1)$-point circular convolution of two length-$(N + M - 1)$ sequences, $g_e[n]$ and $h_e[n]$,

$$y_L[n] = g[n] \, \circledast \, h[n] = g_e[n] \, \circledN \, h_e[n], \qquad (3.12)$$

where $g_e[n]$ and $h_e[n]$ are obtained by *appending* $g[n]$ and $h[n]$ with zero-valued samples:

$$g_e[n] = \begin{cases} g[n], & 0 \le n \le N - 1, \\ 0, & N \le n \le N + M - 1, \end{cases} \qquad (3.13)$$

$$h_e[n] = \begin{cases} h[n], & 0 \le n \le M - 1, \\ 0, & M \le n \le N + M - 1. \end{cases} \qquad (3.14)$$

R3.11   The DFT satisfies a number of useful properties that are often utilized in a number of applications. A detailed listing of these properties and their analytical proofs can be found in any text on digital signal processing. These properties can also be verified using MATLAB. We list below a few selected properties that will be encountered later in this exercise.

Circular Time-Shifting Property – If $G[k]$ denotes the N-point DFT of a length-N sequence $g[n]$, then the N-point DFT of the circularly time-shifted sequence $g[\langle n - n_o \rangle_N]$ is given by $W_N^{kn_o} G[k]$ where $W_N = e^{-j2\pi/N}$.

Circular Frequency-Shifting Property – If $G[k]$ denotes the N-point DFT of a length-N sequence $g[n]$, then the N-point DFT of the sequence $W_N^{-k_o n} g[n]$ is given by $G[\langle k - k_o \rangle_N]$.

Circular Convolution Property – If $G[k]$ and $H[k]$ denote the N-point DFTs of the length-N sequences $g[n]$ and $h[n]$, respectively, then the N-point DFT of the circularly convolved sequence $g[n] \, \circledN \, h[n]$ is given by $G[k]H[k]$.

Parseval's Relation – If $G[k]$ denotes the N-point DFT of a length-N sequence $g[n]$, then

$$\frac{1}{N} \sum_{n=0}^{N-1} |g[n]|^2 = \sum_{k=0}^{N-1} |G[k]|^2. \qquad (3.15)$$

R3.12   The periodic even part $g_{pe}[n]$ and the periodic odd part $g_{po}[n]$ of a length-N real sequence $g[n]$ are given by

$$g_{pe}[n] = \frac{1}{2}\left(g[n] + g[\langle -n \rangle_N]\right), \qquad (3.16)$$

$$g_{po}[n] = \frac{1}{2}\left(g[n] - g[\langle -n \rangle_N]\right). \qquad (3.17)$$

If $G[k]$ denotes the N-point DFT of $g[n]$, then the N-point DFTs of $g_{pe}[n]$ and $g_{po}[n]$ are given by $Re\{G[k]\}$ and $j\,Im\{G[k]\}$, respectively.

R3.13   Let $g[n]$ and $h[n]$ be two length-N real sequences, with $G[k]$ and $H[k]$ denoting their respective N-point DFTs. These two N-point DFTs can be computed efficiently using a single N-point DFT $X[k]$ of a complex length-N sequence $x[n]$ defined by $x[n] = g[n] + jh[n]$ using

$$G[k] = \frac{1}{2}\left(X[k] + X^*[\langle -k \rangle_N]\right), \qquad (3.18)$$

$$H[k] = \frac{1}{2j}\left(X[k] - X^*[\langle -k \rangle_N]\right). \qquad (3.19)$$

R3.14   Let $v[n]$ be a real sequence of length $2N$ with $V[k]$ denoting its $2N$-point DFT. Define two real sequences $g[n]$ and $h[n]$ of length N each as

$$g[n] = v[2n] \ \ and \ \ h[n] = v[2n+1], \qquad 0 \le n < N, \qquad (3.20)$$

with $G[k]$ and $H[k]$ denoting their N-point DFTs. Then the $2N$-point DFT $V[k]$ of $v[n]$ can be computed from the two N-point DFTs, $G[k]$ and $H[k]$, using

$$V[k] = G[\langle k \rangle_N] + W_{2N}^k H[\langle k \rangle_N], \qquad 0 \le k \le 2N - 1. \qquad (3.21)$$

R3.15   The *z-transform* $G(z)$ of a sequence $g[n]$ is defined as

$$G(z) = \mathcal{Z}\{g[n]\} = \sum_{n=-\infty}^{\infty} g[n]z^{-n}, \qquad (3.22)$$

where z is a complex variable. The set $\Re$ of values of z for which the z-transform $G(z)$ converges is called its *region of convergence* (ROC). In general, the region of convergence $\Re$ of a z-transform of a sequence $g[n]$ is an annular region of the z-plane:

$$R_{g-} < |z| < R_{g+}, \qquad (3.23)$$

where $0 \le R_{g-} < R_{g+} \le \infty$.

R3.16   In the case of LTI discrete-time systems , all pertinent z-transforms are rational functions of z^{-1}, that is, they are ratios of two polynomials in z^{-1}:

$$G(z) = \frac{P(z)}{D(z)} = \frac{p_0 + p_1 z^{-1} + \ldots + p_{M-1} z^{-(M-1)} + p_M z^{-M}}{d_0 + d_1 z^{-1} + \ldots + d_{N-1} z^{-(N-1)} + d_N z^{-N}}, \qquad (3.24)$$

which can be alternately written in factored form as

$$G(z) = \frac{p_0}{d_0} \frac{\prod_{r=1}^{M}(1 - \xi_r z^{-1})}{\prod_{s=1}^{N}(1 - \lambda_s z^{-1})} = \frac{p_0}{d_0} z^{N-M} \frac{\prod_{r=1}^{M}(z - \xi_r)}{\prod_{s=1}^{N}(z - \lambda_s)}. \tag{3.25}$$

The *zeros* of $G(z)$ are given by $z = \xi_r$ while the *poles* are given by $z = \lambda_s$. There are additional $(N - M)$ zeros at $z = 0$ (the origin in the z-plane) if $N > M$ or additional $(M - N)$ poles at $z = 0$ if $N < M$.

R3.17   For a sequence with a rational z-transform, the ROC of the z-transform cannot contain any poles and is bounded by the poles.

R3.18   The *inverse z-transform* $g[n]$ of a z-transform $G(z)$ is given by

$$g[n] = \frac{1}{2\pi j} \oint_C G(z) z^{n-1} dz, \tag{3.26}$$

where C is a counterclockwise contour encircling the point $z = 0$ in the ROC of $G(z)$.

R3.19   A rational z-transform $G(z) = P(z)/D(z)$, where the degree of the polynomial $P(z)$ is M and the degree of the polynomial $D(z)$ is N, and with distinct poles at $z = \lambda_s, s = 1, 2, \ldots, N$, can be expressed in a partial-fraction expansion form given by

$$G(z) = \sum_{\ell=0}^{M-N} \eta_\ell z^{-\ell} + \sum_{s=0}^{N} \frac{\rho_s}{1 - \lambda_s z^{-1}}, \tag{3.27}$$

assuming $M \geq N$. The constants ρ_s in the above expression, called the *residues* , are given by

$$\rho_s = (1 - \lambda_s z^{-1})G(z)|_{z=\lambda_s}. \tag{3.28}$$

If $G(z)$ has multiple poles , the partial-fraction expansion is of slightly different form. For example, if the pole at $z = \nu$ is of multiplicity L and the remaining $N - L$ poles are simple and at $z = \lambda_s, s = 1, 2, \ldots, N - L$, then the general partial fraction expansion of $G(z)$ takes the form

$$G(z) = \sum_{\ell=0}^{M-N} \eta_\ell z^{-\ell} + \sum_{s=0}^{N-L} \frac{\rho_s}{1 - \lambda_s z^{-1}} + \sum_{r=1}^{L} \frac{\gamma_r}{(1 - \nu z^{-1})^r}, \tag{3.29}$$

where the constants γ_r (no longer called the residues for $r \neq 1$) are computed using the formula

$$\gamma_r = \frac{1}{(L-r)!(-\nu)^{L-r}} \frac{d^{L-r}}{d(z^{-1})^{L-r}} \left[(1 - \nu z^{-1})^L G(z) \right]_{z=\nu}, \quad r = 1, \ldots, L, \tag{3.30}$$

and the residues ρ_s are calculated using Eq. (3.28).

3.3   MATLAB Commands Used

The MATLAB commands you will encounter in this exercise are as follows:

General Purpose Commands

```
disp
```

Operators and Special Characters

```
:          .          +          -          *          /          ;
%          <          >          .*          ^          .^          ~=
```

Language Constructs and Debugging

```
break      end        error      for        function
if         input      pause
```

Elementary Matrices and Matrix Manipulation

```
fliplr     i        pi        zeros        :
```

Elementary Functions

```
abs        angle      conj       exp        imag       real
rem
```

Polynomial and Interpolation Functions

```
conv
```

Two-Dimensional Graphics

```
axis       grid       plot       stem       title
xlabel     ylabel
```

General Purpose Graphics Functions

```
clf        subplot
```

Character String Functions

```
num2str
```

Data Analysis and Fourier Transform Functions

```
fft        ifft        max       min
```

Signal Processing Toolbox

freqz	impz	residuez	tf2zp	zp2sos
zp2tf	zplane			

For additional information on these commands, see the *MATLAB Reference Guide* [Mat94] and the *Signal Processing Toolbox User's Guide* [Mat96] or type help commandname in the Command window. A brief explanation of the MATLAB functions used here can be found in Appendix B.

3.4   Discrete-Time Fourier Transform

The discrete-time Fourier transform (DTFT) $X(e^{j\omega})$ of a sequence $x[n]$ is a continuous function of ω. Since the data in MATLAB is in vector form, $X(e^{j\omega})$ can only be evaluated at a prescribed set of discrete frequencies. Moreover, only a class of the DTFT that is expressed as a rational function in $e^{-j\omega}$ in the form

$$X(e^{j\omega}) = \frac{p_0 + p_1 e^{-j\omega} + \ldots + p_M e^{-j\omega M}}{d_0 + d_1 e^{-j\omega} + \ldots + d_N e^{-j\omega M}}, \tag{3.31}$$

can be evaluated. In the following two projects you will learn how to evaluate and plot the DTFT and study certain properties of the DTFT using MATLAB.

Project 3.1   DTFT Computation

The DTFT $X(e^{j\omega})$ of a sequence $x[n]$ of the form of Eq. (3.31) can be computed easily at a prescribed set of L discrete frequency points $\omega = \omega_\ell$ using the MATLAB function freqz. Since $X(e^{j\omega})$ is a continuous function of ω, it is necessary to make L as large as possible so that the plot generated using the command plot provides a resonable replica of the actual plot of the DTFT. In MATLAB, freqz computes the L-point DFT of the sequences $\{p_0 \quad p_1 \ldots p_M\}$ and $\{d_0 \quad d_1 \ldots d_M\}$, and then forms their ratio to arrive at $X(e^{j\omega_\ell}), \ell = 1, 2, \ldots, L$. For faster computation, L should be chosen as a power of 2, such as 256 or 512.

Program P3_1 can be used to evaluate and plot the DTFT of the form of Eq. (3.31).

```
% Program P3_1
% Evaluation of the DTFT
clf;
% Compute the frequency samples of the DTFT
w = -4*pi:8*pi/511:4*pi;
num = [2 1];den = [1 -0.6];
h = freqz(num, den, w);
% Plot the DTFT
subplot(2,1,1)
plot(w/pi,real(h));grid
```

```
title('Real part of H(e^{j\omega})')
xlabel('\omega /\pi');
ylabel('Amplitude');
subplot(2,1,2)
plot(w/pi,imag(h));grid
title('Imaginary part of H(e^{j\omega})')
xlabel('\omega /\pi');
ylabel('Amplitude');
pause
subplot(2,1,1)
plot(w/pi,abs(h));grid
title('Magnitude Spectrum |H(e^{j\omega})|')
xlabel('\omega /\pi');
ylabel('Amplitude');
subplot(2,1,2)
plot(w/pi,angle(h));grid
title('Phase Spectrum arg[H(e^{j\omega})]')
xlabel('\omega /\pi');
ylabel('Phase, radians');
```

Questions:

Q3.1 What is the expression of the DTFT being evaluated in Program P3_1? What is the function of the MATLAB command pause?

Q3.2 Run Program P3_1 and compute the real and imaginary parts of the DTFT, and the magnitude and phase spectra . Is the DTFT a periodic function of ω? If it is, what is the period? Explain the type of symmetries exhibited by the four plots.

Q3.3 Modify Program P3_1 to evaluate in the range $0 \leq \omega \leq \pi$ the following DTFT:

$$U(e^{j\omega}) = \frac{0.7 - 0.5e^{-j\omega} + 0.3e^{-j2\omega} + e^{-j3\omega}}{1 + 0.3e^{-j\omega} - 0.5e^{-j2\omega} + 0.7e^{-j3\omega}},$$

and repeat Question Q3.2. Comment on your results. Can you explain the jump in the phase spectrum ? The jump can be removed using the MATLAB command unwrap. Evaluate the phase spectrum with the jump removed.

Q3.4 Modify Program P3_1 to evaluate the DTFT of the following finite-length sequence:

$$g[n] = [1 \quad 3 \quad 5 \quad 7 \quad 9 \quad 11 \quad 13 \quad 15 \quad 17],$$

and repeat Question Q3.2. Comment on your results. Can you explain the jumps in the phase spectrum?

Q3.5 How would you modify Program P3_1 to plot the phase in degrees?

Project 3.2   DTFT Properties

Most of the properties of the DTFT can be verified using MATLAB. In this project you shall verify the properties listed in R3.6. Since all data in MATLAB have to be finite-length vectors, the sequences being used to verify the properties are thus restricted to be of finite length.

Program P3_2 can be used to verify the time-shifting property of the DTFT.

```
% Program P3_2
% Time-Shifting Properties of DTFT
clf;
w = -pi:2*pi/255:pi; wo = 0.4*pi; D = 10;
num = [1 2 3 4 5 6 7 8 9];
h1 = freqz(num, 1, w);
h2 = freqz([zeros(1,D) num], 1, w);
subplot(2,2,1)
plot(w/pi,abs(h1));grid
title('Magnitude Spectrum of Original Sequence')
subplot(2,2,2)
plot(w/pi,abs(h2));grid
title('Magnitude Spectrum of Time-Shifted Sequence')
subplot(2,2,3)
plot(w/pi,angle(h1));grid
title('Phase Spectrum of Original Sequence')
subplot(2,2,4)
plot(w/pi,angle(h2));grid
title('Phase Spectrum of Time-Shifted Sequence')
```

Questions:

Q3.6   Modify Program P3_2 by adding appropriate comment statements and program statements for labeling the two axes of each plot being generated by the program. Which parameter controls the amount of time-shift?

Q3.7   Run the modified program and comment on your results.

Q3.8   Repeat Question Q3.7 for a different value of the time-shift.

Q3.9   Repeat Question Q3.7 for two different sequences of varying lengths and two different time-shifts.

Program P3_3 can be used to verify the frequency-shifting property of the DTFT.

```
% Program P3_3
% Frequency-Shifting Properties of DTFT
clf;
w = -pi:2*pi/255:pi; wo = 0.4*pi;
```

```
num1 = [1 3 5 7 9 11 13 15 17];
L = length(num1);
h1 = freqz(num1, 1, w);
n = 0:L-1;
num2 = exp(wo*i*n).*num1;
h2 = freqz(num2, 1, w);
subplot(2,2,1)
plot(w/pi,abs(h1));grid
title('Magnitude Spectrum of Original Sequence')
subplot(2,2,2)
plot(w/pi,abs(h2));grid
title('Magnitude Spectrum of Frequency-Shifted Sequence')
subplot(2,2,3)
plot(w/pi,angle(h1));grid
title('Phase Spectrum of Original Sequence')
subplot(2,2,4)
plot(w/pi,angle(h2));grid
title('Phase Spectrum of Frequency-Shifted Sequence')
```

Questions:

Q3.10 Modify Program P3_3 by adding appropriate comment statements and program statements for labeling the two axes of each plot being generated by the program. Which parameter controls the amount of frequency-shift?

Q3.11 Run the modified program and comment on your results.

Q3.12 Repeat Question Q3.11 for a different value of the frequency-shift.

Q3.13 Repeat Question Q3.11 for two different sequences of varying lengths and two different frequency-shifts.

Program P3_4 can be used to verify the convolution property of the DTFT.

```
% Program P3_4
% Convolution Property of DTFT
clf;
w = -pi:2*pi/255:pi;
x1 = [1 3 5 7 9 11 13 15 17];
x2 = [1 -2 3 -2 1];
y = conv(x1,x2);
h1 = freqz(x1, 1, w);
h2 = freqz(x2, 1, w);
hp = h1.*h2;
h3 = freqz(y,1,w);
subplot(2,2,1)
plot(w/pi,abs(hp));grid
```

```
title('Product of Magnitude Spectra')
subplot(2,2,2)
plot(w/pi,abs(h3));grid
title('Magnitude Spectrum of Convolved Sequence')
subplot(2,2,3)
plot(w/pi,angle(hp));grid
title('Sum of Phase Spectra')
subplot(2,2,4)
plot(w/pi,angle(h3));grid
title('Phase Spectrum of Convolved Sequence')
```

Questions:

Q3.14 Modify Program P3_4 by adding appropriate comment statements and program statements for labeling the two axes of each plot being generated by the program.

Q3.15 Run the modified program and comment on your results.

Q3.16 Repeat Question Q3.15 for two different sets of sequences of varying lengths.

Program P3_5 can be used to verify the modulation property of the DTFT.

```
% Program P3_5
% Modulation Property of DTFT
clf;
w = -pi:2*pi/255:pi;
x1 = [1 3 5 7 9 11 13 15 17];
x2 = [1 -1 1 -1 1 -1 1 -1 1];
y = x1.*x2;
h1 = freqz(x1, 1, w);
h2 = freqz(x2, 1, w);
h3 = freqz(y,1,w);
subplot(3,1,1)
plot(w/pi,abs(h1));grid
title('Magnitude Spectrum of First Sequence')
subplot(3,1,2)
plot(w/pi,abs(h2));grid
title('Magnitude Spectrum of Second Sequence')
subplot(3,1,3)
plot(w/pi,abs(h3));grid
title('Magnitude Spectrum of Product Sequence')
```

Questions:

Q3.17 Modify Program P3_5 by adding appropriate comment statements and program statements for labeling the two axes of each plot being generated by the program.

Q3.18 Run the modified program and comment on your results.

Q3.19   Repeat Question Q3.18 for two different sets of sequences of varying lengths.

Program P3_6 can be used to verify the time-reversal property of the DTFT.

```
% Program P3_6
% Time-Reversal Property of DTFT
clf;
w = -pi:2*pi/255:pi;
num = [1 2 3 4];
L = length(num)-1;
h1 = freqz(num, 1, w);
h2 = freqz(fliplr(num), 1, w);
h3 = exp(w*L*i).*h2;
subplot(2,2,1)
plot(w/pi,abs(h1));grid
title('Magnitude Spectrum of Original Sequence')
subplot(2,2,2)
plot(w/pi,abs(h3));grid
title('Magnitude Spectrum of Time-Reversed Sequence')
subplot(2,2,3)
plot(w/pi,angle(h1));grid
title('Phase Spectrum of Original Sequence')
subplot(2,2,4)
plot(w/pi,angle(h3));grid
title('Phase Spectrum of Time-Reversed Sequence')
```

Questions:

Q3.20   Modify Program P3_6 by adding appropriate comment statements and program statements for labeling the two axes of each plot being generated by the program. Explain how the program implements the time-reversal operation.

Q3.21   Run the modified program and comment on your results.

Q3.22   Repeat Question Q3.21 for two different sequences of varying lengths.

3.5   Discrete Fourier Transform

The discrete Fourier transform (DFT) $X[k]$ of a finite-length sequence $x[n]$ can be easily computed in MATLAB using the function `fft`. There are two versions of this function. `fft(x)` computes the DFT $X[k]$ of the sequence $x[n]$ where the length of $X[k]$ is the same as that of $x[n]$. `fft(x,L)` computes the L-point DFT of a sequence $x[n]$ of length N where $L \geq N$. If $L > N$, $x[n]$ is zero-padded with $L - N$ trailing zero-valued samples before the DFT is computed. The inverse discrete Fourier transform (IDFT) $x[n]$ of a DFT sequence $X[k]$ can likewise be computed using the function `ifft`, which also has two versions.

Project 3.3 DFT and IDFT Computations

Questions:

Q3.23 Write a MATLAB program to compute and plot the L-point DFT $X[k]$ of a sequence $x[n]$ of length N with $L \geq N$ and then to compute and plot the L-point IDFT of $X[k]$. Run the program for sequences of different lengths N and for different values of the DFT length L. Comment on your results.

Q3.24 Write a MATLAB program to compute the N-point DFT of two length-N real sequences using a single N-point DFT and compare the result by computing directly the two N-point DFTs (see R3.13).

Q3.25 Write a MATLAB program to compute the $2N$-point DFT of a length-$2N$ real sequence using a single N-point DFT and compare the result by computing directly the $2N$-point DFT (see R3.14).

Project 3.4 DFT Properties

Two important concepts used in the application of the DFT are the *circular-shift* of a sequence and the *circular convolution* of two sequences of the same length. As these operations are needed in verifying certain properties of the DFT, we implement them as MATLAB functions circshift and circonv as indicated below:

```
function y = circshift(x,M)
% Develops a sequence y obtained by
% circularly shifting a finite-length
% sequence x by M samples
if abs(M) > length(x)
    M = rem(M,length(x));
end
if M < 0
    M = M + length(x);
end
y = [x(M+1:length(x)) x(1:M)];

function y = circonv(x1,x2)
L1 = length(x1); L2 = length(x2);
if L1 ~= L2, error('Sequences of unequal lengths'), end
y = zeros(1,L1);
x2tr = [x2(1) x2(L2:-1:2)];
for k = 1:L1
    sh = circshift(x2tr,1-k);
    h = x1.*sh;
    y(k) = sum(h);
end
```

Questions:

Q3.26   What is the purpose of the command `rem` in the function `circshift`?

Q3.27   Explain how the function `circshift` implements the circular time-shifting operation.

Q3.28   What is the purpose of the operator `~=` in the function `circonv`?

Q3.29   Explain how the operation of the function `circonv` implements the circular convolution operation.

Program P3_7 can be used to illustrate the concept of circular shift of a finite-length sequence. It employs the function `circshift`.

```
% Program P3_7
% Illustration of Circular Shift of a Sequence
clf;
M = 6;
a = [0 1 2 3 4 5 6 7 8 9];
b = circshift(a,M);
L = length(a)-1;
n = 0:L;
subplot(2,1,1);
stem(n,a);axis([0,L,min(a),max(a)]);
title('Original Sequence');
subplot(2,1,2);
stem(n,b);axis([0,L,min(a),max(a)]);
title(['Sequence Obtained by Circularly Shifting by ',num2str(M),'
Samples']);
```

Questions:

Q3.30   Modify Program P3_7 by adding appropriate comment statements and program statements for labeling each plot being generated by the program. Which parameter determines the amount of time-shifting? What happens if the amount of time-shift is greater than the sequence length?

Q3.31   Run the modified program and verify the circular time-shifting operation.

Program P3_8 can be used to illustrate the circular time-shifting property of the DFT. It employs the function `circshift`.

```
% Program P3_8
% Circular Time-Shifting Property of DFT
clf;
x = [0 2 4 6 8 10 12 14 16];
N = length(x)-1; n = 0:N;
```

```
y = circshift(x,5);
XF = fft(x);
YF = fft(y);
subplot(2,2,1)
stem(n,abs(XF)); grid
title('Magnitude of DFT of Original Sequence');
subplot(2,2,2)
stem(n,abs(YF)); grid
title('Magnitude of DFT of Circularly Shifted Sequence');
subplot(2,2,3)
stem(n,angle(XF)); grid
title('Phase of DFT of Original Sequence');
subplot(2,2,4)
stem(n,angle(YF)); grid
title('Phase of DFT of Circularly Shifted Sequence');
```

Questions:

Q3.32 Modify Program P3_8 by adding appropriate comment statements and program statements for labeling each plot being generated by the program. What is the the amount of time-shift?

Q3.33 Run the modified program and verify the circular time-shifting property of the DFT.

Q3.34 Repeat Question Q3.33 for two different amounts of time-shift.

Q3.35 Repeat Question Q3.33 for two different sequences of different lengths.

Program P3_9 can be used to illustrate the circular convolution property of the DFT. It employs the function circonv.

```
% Program P3_9
% Circular Convolution Property of DFT
g1 = [1 2 3 4 5 6]; g2 = [1 -2 3 3 -2 1];
ycir = circonv(g1,g2);
disp('Result of circular convolution = ');disp(ycir)
G1 = fft(g1); G2 = fft(g2);
yc = real(ifft(G1.*G2));
disp('Result of IDFT of the DFT products = ');disp(yc)
```

Questions:

Q3.36 Run Program P3_9 and verify the circular convolution property of the DFT.

Q3.37 Repeat Question Q3.36 for two other different sets of equal-length sequences.

Program P3_10 can be used to illustrate the relation between circular and linear convolutions (see R3.10).

```
% Program P3_10
% Linear Convolution via Circular Convolution
g1 = [1 2 3 4 5];g2 = [2 2 0 1 1];
g1e = [g1 zeros(1,length(g2)-1)];
g2e = [g2 zeros(1,length(g1)-1)];
ylin = circonv(g1e,g2e);
disp('Linear convolution via circular convolution = ');disp(ylin);
y = conv(g1, g2);
disp('Direct linear convolution = ');disp(y)
```

Questions:

Q3.38   Run Program P3_10 and verify that linear convolution can be obtained via circular convolution.

Q3.39   Repeat Question Q3.38 for two other different sets of sequences of unequal lengths.

Q3.40   Write a MATLAB program to develop the linear convolution of two sequences via the DFT of each. Using this program verify the results of Questions Q3.38 and Q3.39.

Program P3_11 can be used to verify the relation between the DFTs of the periodic even and the periodic odd parts of a real sequence, and its DFT (see R3.12).

```
% Program P3_11
% Relations between the DFTs of the Periodic Even
% and Odd Parts of a Real Sequence
x = [1 2 4 2 6 32 6 4 2 zeros(1,247)];
x1 = [x(1) x(256:-1:2)];
xe = 0.5 *(x + x1);
XF = fft(x);
XEF = fft(xe);
clf;
k = 0:255;
subplot(2,2,1);
plot(k/128,real(XF)); grid
ylabel('Amplitude');
title('Re(DFT\{x[n]\})');
subplot(2,2,2);
plot(k/128,imag(XF)); grid
ylabel('Amplitude');
title('Im(DFT\{x[n]\})');
subplot(2,2,3);
plot(k/128,real(XEF)); grid
xlabel('Time index n');ylabel('Amplitude');
```

```
title('Re(DFT\{x_{e}[n]\})');
subplot(2,2,4);
plot(k/128,imag(XEF)); grid
xlabel('Time index n');ylabel('Amplitude');
title('Im(DFT\{x_{e}[n]\})');
```

Questions:

Q3.41 What is the relation between the sequences x1[n] and x[n]?

Q3.42 Run Program P3_11. The imaginary part of XEF should be zero as the DFT of the periodic even part is simply the real part of XEF of the original sequence. Can you verify that? How can you explain the simulation result?

Q3.43 Modify the program to verify the relation between the DFT of the periodic odd part and the imaginary part of XEF.

Parseval's relation (Eq. (3.15)) can be verified using the following program.

```
% Program P3_12
% Parseval's Relation
x = [(1:128) (128:-1:1)];
XF = fft(x);
a = sum(x.*x)
b = round(sum(abs(XF).^2)/256);
```

Questions:

Q3.44 Run Program P3_12. Do you get the same values for a and b?

Q3.45 Modify the program in such a way that you do not have to use the command abs(XF). Use the MATLAB command conj(x) to compute the complex conjugate of x.

3.6 z-Transform

As in the case of the discrete-time Fourier transform, we restrict our attention here to a z-transform $G(z)$ of a sequence $g[n]$ that is a rational function of the complex variable z^{-1} and expressed in the form of a ratio of polynomials in z^{-1} as in Eq. (3.24) or in factored form as in Eq. (3.25). Some of the operations that are of interest in practice are as follows. (1) Evaluate the z-transform $G(z)$ on the unit circle, that is, evaluate $G(e^{j\omega})$; (2) develop the pole-zero plot of $G(z)$; (3) develop the factored form of $G(z)$; (4) determine the inverse z-transform $g[n]$ of $G(z)$; and (5) make a partial-fraction expansion of $G(z)$. In the next two projects you will learn how to perform the above operations using MATLAB.

Project 3.5   Analysis of z-Transforms

The function freqz can be used to evaluate the values of a rational z-transform on the unit circle . To this end, Program P3_1 can be used without any modifications.

Question:

Q3.46 Using Program P3_1 evaluate the following z-transform on the unit circle:

$$G(z) = \frac{2 + 5z^{-1} + 9z^{-2} + 5z^{-3} + 3z^{-4}}{5 + 45z^{-1} + 2z^{-2} + z^{-3} + z^{-4}}. \tag{3.32}$$

The pole-zero plot of a rational z-transform $G(z)$ can be readily obtained using the function zplane. There are two versions of this function. If the z-transform is given in the form of a rational function as in Eq. (3.32), the command to use is zplane(num, den) where num and den are row vectors containing the coefficients of the numerator and denominator polynomials of $G(z)$ in ascending powers of z^{-1}. On the other hand, if the zeros and poles of $G(z)$ are given, the command to use is zplane(zeros, poles) where zeros and poles are column vectors. In the pole-zero plot generated by MATLAB, the location of a pole is indicated by the symbol $\times$ and the location of a zero is indicated by the symbol o.

The function tf2zp can be used to determine the zeros and poles of a rational z-transform $G(z)$. The program statement to use is [z, p, k] = tf2zp(num,den) where num and den are row vectors containing the coefficients of the numerator and denominator polynomials of $G(z)$ in ascending powers of z^{-1} and the output file contains the gain constant k and the computed zeros and poles given as column vectors z and p, respectively. The factored form of the z-transform can be obtained from the zero-pole description using the function sos = zp2sos(z,p,k). The function computes the coefficients of each second-order factor given as an $L \times 6$ matrix sos where

$$\text{sos} = \begin{bmatrix} b_{01} & b_{11} & b_{21} & a_{01} & a_{11} & a_{21} \\ b_{02} & b_{12} & b_{22} & a_{02} & a_{12} & a_{22} \\ \vdots & \vdots & \vdots & \vdots & \vdots & \vdots \\ b_{0L} & b_{1L} & b_{2L} & a_{0L} & a_{1L} & a_{2L} \end{bmatrix},$$

where the ℓth row contains the coefficients of the numerator and the denominator of the ℓth second order factor of the z-transform $G(z)$:

$$G(z) = \prod_{\ell=1}^{L} \frac{b_{0\ell} + b_{1\ell}\, z^{-1} + b_{2\ell}\, z^{-2}}{a_{0\ell} + a_{1\ell}\, z^{-1} + a_{2\ell}\, z^{-2}}.$$

Questions:

Q3.47 Write a MATLAB program to compute and display the poles and zeros, to compute and display the factored form, and to generate the pole-zero plot of a z-transform that is a ratio of two polynomials in z^{-1}. Using this program, analyze the z-transform $G(z)$ of Eq. (3.32).

Q3.48 From the pole-zero plot generated in Question Q3.47, determine the number of regions of convergence (ROC) of $G(z)$. Show explicitly all possible ROCs . Can you tell from the pole-zero plot whether or not the DTFT exists?

The reverse process of converting a z-transform given in the form of zeros, poles, and the gain constant to a rational form can be implemented using the function zp2tf. The program statement to use is [num,den] = zp2tf(z,p,k).

Question:

Q3.49 Write a MATLAB program to compute and display the rational z-transform from its zeros, poles and gain constant. Using this program, determine the rational form of a z-transform whose zeros are at $\xi_1 = 0.3, \xi_2 = 2.5, \xi_3 = -0.2 + j\,0.4$, and $\xi_4 = -0.2 - j\,0.4$; the poles are at $\lambda_1 = 0.5, \lambda_2 = -0.75, \lambda_3 = 0.6 + j\,0.7$, and $\lambda_4 = 0.6 - j\,0.7$; and the gain constant k is 3.9.

Project 3.6   Inverse z-Transform

The inverse $g[n]$ of a rational z-transform $G(z)$ can be computed using MATLAB in basically two different ways . To this end, it is necessary to know a priori the ROC of $G(z)$. The function impz provides the samples of the time-domain sequence, which is assumed to be causal. There are three versions of this function: [g,t] = impz(num,den), [g,t] = impz(num,den, L), and [g,t] = impz(num,den, L, FT), where num and den are row vectors containing the coefficients of the numerator and denominator polynomials of $G(z)$ in ascending powers of z^{-1}, L is the desired number of the samples of the inverse transform, g is the vector containing the samples of the inverse transform starting with the sample at $n = 0$, t is the length of g, and FT is the specified sampling frequency in Hz with default value of unity.

A closed-form expression for the inverse of a rational z-transform can be obtained by first performing a partial-fraction expansion using the function residuez and then determining the inverse of each term in the expansion by looking up a table of z-transforms. The function residuez can also be used to convert a z-transform given in the form of a partial-fraction expansion to a ratio of polynomials in z^{-1}.

Questions:

Q3.50 Write a MATLAB program to compute the first L samples of the inverse of a rational z-transform where the value of L is provided by the user through the command input. Using this program compute and plot the first 50 samples of the inverse of $G(z)$ of Eq. (3.32). Use the command stem for plotting the sequence generated by the inverse transform.

Q3.51 Write a MATLAB program to determine the partial-fraction expansion of a rational z-transform. Using this program determine the partial-fraction expansion of $G(z)$ of Eq. (3.32) and then its inverse z-transform $g[n]$ in closed form. Assume $g[n]$ to be a causal sequence.

3.7   Background Reading

[1] A. Antoniou. *Digital Filters: Analysis, Design, and Applications.* McGraw-Hill, New York NY, 1993, second edition. Chs. 2, 13.

[2] E. Cunningham. *Digital Filtering: An Introduction.* Houghton-Mifflin, Boston, MA, 1992. Ch. 3.

[3] D.J. DeFatta, J.G. Lucas, and W.S. Hodgkiss. *Digital Signal Processing: A System Design Approach.* Wiley, New York, NY, 1988. Secs. 2.1, 3.1–3.3, 6.1–6.4.

[4] L.B. Jackson, *Digital Filters and Signal Processing.* Kluwer, Boston MA, third edition, 1996. Ch. 3 and Secs. 6.1, 6.2, 7.1, 7.2.

[5] R. Kuc. *Introduction to Digital Signal Processing.* McGraw-Hill, New York NY, 1988. Chs. 3–5.

[6] L.C. Ludeman. *Fundamentals of Digital Signal Processing.* Harper & Row, New York NY, 1986. Secs. 1.4, 2.1, 2.2, 6.3.

[7] S.K. Mitra. *Digital Signal Processing: A Computer-Based Approach.* McGraw-Hill, New York NY, 1998. Ch. 3.

[8] A.V. Oppenheim and R.W. Schafer. *Discrete-Time Signal Processing.* Prentice-Hall, Englewood Cliffs NJ, 1989. Secs. 2.6–2.9 and Chs. 4, 8.

[9] S.J. Orfanidis. *Introduction to Signal Processing.* Prentice-Hall, Englewood Cliffs NJ, 1996. Ch. 5 and Secs. 9.1, 9.2.

[10] B. Porat. *A Course in Digital Signal Procesing.* Wiley, New York NY, 1996. Secs. 2.7, 4.1–4.7, 7.1–7.6.

[11] J.G. Proakis and D.G. Manolakis. *Digital Signal Processing: Principles, Algorithms, and Applications.* Prentice-Hall, Englewood Cliffs NJ, third edition, 1996. Secs. 1.3, 3.1–3.4, 4.2, 4.3, 5.1, 5.2.

[12] R.A. Roberts and C.T. Mullis. *Digital Signal Processing.* Addison-Wesley, Reading MA, 1987. Chs. 3, 4.

LTI Discrete-Time Systems in the Frequency-Domain

4

4.1   Introduction

A linear, time-invariant (LTI) discrete-time system is completely characterized in the time-domain by its impulse response sequence, and the output sequence of the LTI system can be computed for any input sequence by convolving the input sequence with its impulse response sequence. Certain classes of LTI discrete-time systems are characterized also by a linear, constant-coefficient difference equation. For such systems, the output sequence can be computed recursively for any input sequence. By applying the DTFT or the z-transform to either the convolution sum description or to the difference equation representation, the LTI discrete-time system can also be characterized in the frequency domain. Such transform domain representations provide additional insight into the behavior of such systems, in addition to making it simpler to design and implement them for specific applications.

4.2   Background Review

R4.1   If $\{h[n]\}$ denotes the impulse response sequence of an LTI discrete-time system, its *frequency response* $H(e^{j\omega})$ is given by the discrete-time Fourier transform of $\{h[n]\}$, that is,

$$H(e^{j\omega}) = \sum_{n=-\infty}^{\infty} h[n]\, e^{-j\omega n}. \tag{4.1}$$

R4.2   In general, $H(e^{j\omega})$ is a complex function of ω with a period 2π and can be expressed in terms of its real and imaginary parts or its magnitude and phase. Thus,

$$H(e^{j\omega}) = H_{re}(e^{j\omega}) + j\, H_{im}(e^{j\omega}) = |H(e^{j\omega})|\, e^{j\,\theta(\omega)}, \tag{4.2}$$

where $H_{re}(e^{j\omega})$ and $H_{im}(e^{j\omega})$ are, respectively, the real and imaginary parts of $H(e^{j\omega})$, and

$$\theta(\omega) = arg\{H(e^{j\omega})\}. \tag{4.3}$$

The quantity $|H(e^{j\omega})|$ is called the *magnitude response* and the quantity $\theta(\omega)$ is called the *phase response* of the LTI discrete-time system.

R4.3   The *gain function* $\mathcal{G}(\omega)$ of the LTI system is defined by

$$\mathcal{G}(\omega) = 20 \log_{10} |H(e^{j\omega})| \quad \text{dB}. \tag{4.4}$$

53

The negative of the gain function, $a(\omega) = -\mathcal{G}(\omega)$, is called the *attenuation* or *loss function*.

R4.4   For a discrete-time system characterized by a real impulse response $h[n]$, the magnitude function is an even function of ω; that is, $|H(e^{j\omega})| = |H(e^{-j\omega})|$, and the phase function is an odd function of ω; that is, $\theta(\omega) = -\theta(-\omega)$. Likewise, $H_{re}(e^{j\omega})$ is an even function of ω and $H_{im}(e^{j\omega})$ is an odd function of ω.

R4.5   The phase responses of discrete-time systems when determined by a computer may exhibit jumps by an amount of 2π caused by the way the arctangent function is computed. The phase response can be made a continuous function of ω by unwrapping the phase response across the jumps by adding multiples of $\pm 2\pi$.

R4.6   The *group delay function* of an LTI discrete-time system is defined by

$$\tau(\omega) = -\frac{d\,\theta_c(\omega)}{d\omega},\tag{4.5}$$

where $\theta_c(\omega)$ denotes the unwrapped phase function. If the phase function is in radians, then the group delay is in seconds.

R4.7   The steady-state output $y[n]$ of a real coefficient LTI discrete-time system with a frequency response $H(e^{j\omega})$ for an input

$$x[n] = A\,\cos(\omega_o n + \phi),\tag{4.6}$$

with A real, is given by

$$y[n] = A\,|H(e^{j\omega}_o)|\,\cos(\omega_o n + \theta(\omega_o) + \phi).\tag{4.7}$$

R4.8   From the convolution sum description of an LTI discrete-time system as given by Eq. (2.4), it follows that the frequency response of an LTI discrete-time system is given by the ratio of the Fourier transform $Y(e^{j\omega})$ of the output sequence $y[n]$ to the Fourier transform $X(e^{j\omega})$ of the input sequence $x[n]$, that is,

$$H(e^{j\omega}) = Y(e^{j\omega})/X(e^{j\omega}).\tag{4.8}$$

R4.9   For an LTI system characterized by a linear constant-coefficient difference equation of the form of Eq. (2.11), the frequency response $H(e^{j\omega})$ can be expressed as

$$H(e^{j\omega}) = \frac{\sum_{k=0}^{M} p_k e^{-j\omega k}}{\sum_{k=0}^{N} d_k e^{-j\omega k}}.\tag{4.9}$$

R4.10   The z-transform $H(z)$ of the impulse response sequence $\{h[n]\}$ of the LTI discrete-time system is called the *transfer function* or the *system function* . From the convolution sum description of an LTI discrete-time system as given by Eq. (2.4) it follows that the transfer function $H(z)$ of an LTI discrete-time system is given by the ratio of the z-transform $Y(z)$ of the output sequence $y[n]$ to the z-transform $X(z)$ of the input sequence $x[n]$; that is, $H(z) = Y(z)/X(z)$.

R4.11 If the ROC of $H(z)$ includes the unit circle, it is then related to the frequency response $H(e^{j\omega})$ of the LTI discrete-time system through

$$H(e^{j\omega}) = H(z)|_{z=e^{j\omega}}. \tag{4.10}$$

R4.12 For a real-coefficient transfer function $H(z)$:

$$|H(e^{j\omega})|^2 = H(e^{j\omega})\,H^*(e^{j\omega}) = H(e^{j\omega})\,H(e^{-j\omega}) = H(z)\,H(z^{-1})|_{z=e^{j\omega}}. \tag{4.11}$$

R4.13 For an LTI system characterized by a linear constant-coefficient difference equation of the form of Eq. (2.11), the transfer function $H(z)$ can be expressed as :

$$H(z) = \frac{Y(z)}{X(z)} = \frac{p_0 + p_1\,z^{-1} + \ldots + p_M\,z^{-M}}{d_0 + d_1\,z^{-1} + \ldots + d_N\,z^{-N}}. \tag{4.12}$$

R4.14 The transfer function of Eq. (4.12) can also be expressed in the form

$$H(z) = \frac{p_0 \prod_{k=1}^{M}(1 - \xi_k\,z^{-1})}{d_0 \prod_{k=1}^{N}(1 - \lambda_k\,z^{-1})}, \tag{4.13}$$

where $\xi_1, \xi_2, \ldots, \xi_M$ are the finite zeros and $\lambda_1, \lambda_2, \ldots, \lambda_N$ are the finite poles of $H(z)$. If $N > M$, there are additional $(N - M)$ zeros at $z = 0$, and if $N < M$, there are additional $(M - N)$ poles at $z = 0$.

R4.15 All poles of a stable causal transfer function $H(z)$ must be strictly inside the unit circle.

R4.16 The frequency responses of the four popular types of ideal zero-phase digital filters with real impulse response coefficients are shown in Figure 4.1. An ideal filter has a magnitude response equal to unity in the passband and to zero in the stop band, and has a zero phase everywhere.

R4.17 The impulse response $h_{LP}[n]$ of the ideal lowpass filter of Figure 4.1 is given by

$$h_{LP}[n] = \frac{\sin(\omega_c n)}{\pi n}, \quad -\infty < n < \infty. \tag{4.14}$$

R4.18 A first-order lowpass IIR transfer function $H_{LP}(z)$ is given by

$$H_{LP}(z) = \frac{1 - \alpha}{2} \cdot \frac{1 + z^{-1}}{1 - \alpha z^{-1}}, \tag{4.15}$$

where $|\alpha| < 1$ for stability. The frequency ω_c where the gain is 3 dB below its maximum value at dc ($\omega = 0$), called the *3-dB cutoff frequency* , is related to the parameter α through

$$\alpha = \frac{1 - \sin \omega_c}{\cos \omega_c}. \tag{4.16}$$

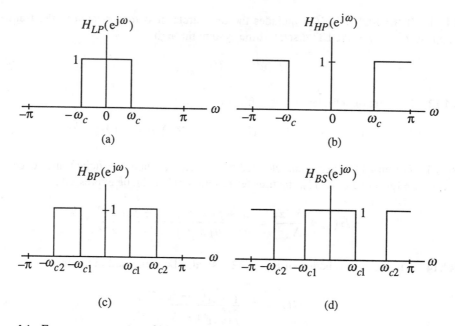

Figure 4.1 Frequency responses of ideal filters: (a) lowpass filter, (b) highpass filter, (c) bandpass filter, and (d) bandstop filter.

R4.19 A first-order highpass IIR transfer function $H_{HP}(z)$ is given by

$$H_{HP}(z) = \frac{1+\alpha}{2} \cdot \frac{1-z^{-1}}{1-\alpha z^{-1}}, \tag{4.17}$$

where $|\alpha| < 1$ for stability. Its 3-dB cutoff frequency ω_c is also given by Eq. (4.16).

R4.20 A second-order bandpass IIR transfer function $H_{BP}(z)$ is given by

$$H_{BP}(z) = \frac{1-\alpha}{2} \cdot \frac{1-z^{-2}}{1-\beta(1+\alpha)z^{-1}+\alpha z^{-2}}. \tag{4.18}$$

Its magnitude response goes to zero at $\omega = 0$ and at $\omega = \pi$ and assumes a maximum value of unity at $\omega = \omega_o$, called the *center frequency* of the bandpass filter, where

$$\omega_o = \cos^{-1}(\beta). \tag{4.19}$$

The *3-dB bandwidth* $\Delta\omega_{3dB}$, given by the difference of the two 3-dB cutoff frequencies, is given by

$$\Delta\omega_{3dB} = \omega_{c2} - \omega_{c1} = \cos^{-1}\left(\frac{2\alpha}{1+\alpha^2}\right). \tag{4.20}$$

R4.21 A second-order bandstop IIR transfer function $H_{BS}(z)$ is given by

$$H_{BS}(z) = \frac{1+\alpha}{2} \cdot \frac{1-2\beta z^{-1}+z^{-2}}{1-\beta(1+\alpha)z^{-1}+\alpha z^{-2}}. \tag{4.21}$$

Its magnitude response takes the maximum value of unity at $\omega = 0$ and at $\omega = \pi$ and goes to zero at $\omega = \omega_o$, called the *notch frequency* , where ω_o is given by Eq. (4.19). The *3-dB notch bandwidth* $\Delta\omega_{3dB}$ is given by Eq. (4.20).

R4.22 By cascading the simple digital filters described above, digital filters with sharper magnitude responses can be implemented. For example, for a cascade of K first-order lowpass sections characterized by the transfer function of Eq. (4.15), the overall structure has a transfer function $G_{LP}(z)$ given by

$$G_{LP}(z) = \left(\frac{1-\alpha}{2} \cdot \frac{1+z^{-1}}{1-\alpha z^{-1}} \right)^K . \qquad (4.22)$$

The parameters α and K are related to the 3-dB cutoff frequency ω_c of the cascade through

$$\alpha = \frac{1 + (1-B)\cos \omega_c - \sin \omega_c \sqrt{2B - B^2}}{1 - B + \cos \omega_c}, \qquad (4.23)$$

where

$$B = 2^{(K-1)/K}. \qquad (4.24)$$

R4.23 For non–real time processing of real input signals of finite length, zero-phase filtering can be very simply implemented if the causality requirement is relaxed. In one scheme, the finite-length input data are processed through a causal real-coefficient filter $H(z)$ whose output is then time-reversed and processed by the same filter once again as indicated in Figure 4.2.

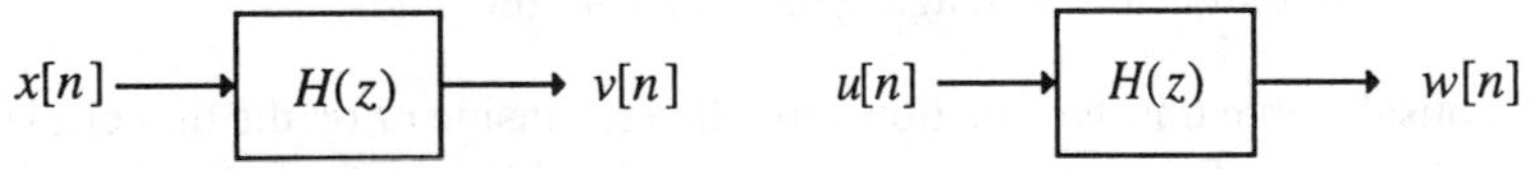

$$u[n] = v[-n], \quad y[n] = w[-n]$$

Figure 4.2 Implementation of a zero-phase filtering scheme.

R4.24 It is always possible to design an FIR transfer function with an exact linear phase response. Such a transfer function corresponds either to a symmetric impulse response defined by

$$h[n] = h[N - n], \quad 0 \le n \le N, \qquad (4.25)$$

or an antisymmetric impulse response defined by

$$h[n] = -h[N - n], \quad 0 \le n \le N, \qquad (4.26)$$

where N is the order of the transfer function and the length of $h[n]$ is $N + 1$. There are four types of such transfer functions:

Type 1:   Symmetric Impulse Response with Odd Length.
Type 2:   Symmetric Impulse Response with Even Length.
Type 3:   Antisymmetric Impulse Response with Odd Length.
Type 4:   Antisymmetric Impulse Response with Even Length.

R4.25   A Type 2 FIR transfer function must have a zero at $z = -1$, and as a result, it cannot be used to design a highpass filter. A Type 3 FIR transfer function must have a zero at $z = 1$ and $z = -1$ and, therefore, cannot be used to design either a lowpass, a highpass, or a bandstop filter. A Type 4 FIR transfer function is not appropriate for designing a lowpass filter due to the presence of a zero at $z = 1$. The Type 1 FIR filter has no such restrictions and can be used to design almost any type of filter.

R4.26   A causal stable real coefficient transfer function $H(z)$ is defined as a *bounded real* (BR) transfer function if

$$|H(e^{j\omega})| \leq 1 \qquad \text{for all } \omega. \tag{4.27}$$

R4.27   A transfer function $A(z)$ with unity magnitude response for all frequencies, that is,

$$|A(e^{j\omega})|^2 = 1 \qquad \text{for all } \omega, \tag{4.28}$$

is called an *allpass* transfer function. An M-th order causal real-coefficient IIR allpass transfer function is of the form

$$A(z) = \pm \frac{z^{-M} D_M(z^{-1})}{D_M(z)}, \tag{4.29}$$

where $D_M(z)$ is a polynomial of degree M. The poles and the zeros of a real-coefficient allpass function exhibit *mirror-image symmetry* in the z-plane. If the allpass transfer function is also causal and stable, then all its poles are inside the unit circle and all its zeros are outside the unit circle in a mirror-image symmetry with the poles.

R4.28   A causal stable transfer function with all zeros inside or on the unit circle is called a *minimum-phase transfer function*, whereas a causal stable transfer function with all zeros outside the unit circle is called a *maximum-phase transfer function*.

R4.29   A set of M transfer functions $\{H_0(z), H_1(z), \ldots, H_{M-1}(z)\}$ is defined to be *delay-complementary* of each other, if the sum of their transfer functions is equal to some integer multiple of the unit delay, that is,

$$\sum_{k=0}^{M-1} H_k(z) = \beta z^{-n_o}, \qquad \beta \neq 0, \tag{4.30}$$

where n_o is a nonnegative integer. The delay-complementary transfer function $H_1(z)$ to a Type 1 linear-phase FIR transfer function $H_0(z)$ of odd length L is simply given by $H_1(z) = z^{-(L-1)/2} - H_0(z)$.

R4.30   A set of M digital filters $\{H_i(z)\}, i = 0, 1, \ldots, M - 1$, are defined to be *allpass-complementary* of each other if the sum of their transfer functions is equal to an allpass function $A(z)$, that is,

$$\sum_{i=0}^{M-1} H_i(z) = A(z). \tag{4.31}$$

R4.31 A set of M digital filters $\{H_i(z)\}$, $i = 0, 1, \ldots, M - 1$, are defined to be *power-complementary* of each other if the sum of the squares of their magnitude responses is equal to one, that is,

$$\sum_{i=0}^{M-1} |H_i(e^{j\omega})|^2 = 1 \qquad \text{for all } \omega. \qquad (4.32)$$

R4.32 Let $A_m(z)$ be a real-coefficient allpass function of mth order:

$$A_m(z) = \frac{d_m + d_{m-1} z^{-1} + d_{m-2} z^{-2} + \ldots + d_1 z^{-(m-1)} + z^{-m}}{1 + d_1 z^{-1} + d_2 z^{-2} + \ldots + d_{m-1} z^{-(m-1)} + d_m z^{-m}}. \qquad (4.33)$$

Generate an $(m - 1)$th order real-coefficient allpass function $A_{m-1}(z)$ according to

$$
\begin{aligned}
A_{m-1}(z) &= z \left[\frac{A_m(z) - k_m}{1 - k_m A_m(z)} \right] \\
&= \frac{d'_{m-1} + d'_{m-2} z^{-1} + \ldots + d'_1 z^{-(m-2)} + z^{-(m-1)}}{1 + d'_1 z^{-1} + \ldots + d'_{m-2} z^{-(m-2)} + d'_{m-1} z^{-(m-1)}}, \qquad (4.34)
\end{aligned}
$$

where

$$d'_i = \frac{d_i - d_m d_{m-i}}{1 - d_m^2}, \qquad i = 1, 2, \ldots, m - 1. \qquad (4.35)$$

Define $k_m = A_m(\infty)$. The necessary and sufficient conditions for $A_m(z)$ to be stable are (1) $k_m^2 < 1$ and (2) $A_{m-1}(z)$ is a stable allpass function. The process can be continued to test the stability of $A_{m-1}(z)$ by generating an $(m - 2)$th order allpass function, and so on, resulting in a set of allpass functions of decreasing orders

$$A_m(z), \quad A_{m-1}(z), \quad \ldots, \quad A_2(z), \quad A_1(z), \quad A_0(z) = 1,$$

and a set of coefficients

$$k_m, \quad k_{m-1}, \quad \ldots, \quad k_2, \quad k_1.$$

The allpass function $A_m(z)$ is stable if and only if $k_\ell^2 < 1$ for $\ell = m, m - 1, \ldots, 1$.

4.3 MATLAB Commands Used

The MATLAB commands you will encounter in this exercise are as follows:

General Purpose Commands

```
disp
```

Operators and Special Characters

```
    :          .           +         -         *         /         ;
    %
```

Language Constructs and Debugging

```
function        pause
```

Elementary Matrices and Matrix Manipulation

```
fliplr        pi
```

Elementary Functions

```
abs        angle        imag        log10        real
```

Two-Dimensional Graphics

```
axis        grid        plot        stem        title
xlabel       ylabel
```

General Purpose Graphics Functions

```
clf        subplot
```

Signal Processing Toolbox

```
filter        filtfilt        freqz        grpdelay        impz
poly2rc       sinc            zplane
```

For additional information on these commands, see the *MATLAB Reference Guide* [Mat94] and the *Signal Processing Toolbox User's Guide* [Mat96] or type help commandname in the Command window. A brief explanation of the MATLAB functions used here can be found in Appendix B.

4.4   Transfer Function and Frequency Response

The z-transform of the impulse response sequence $\{h[n]\}$ of an LTI discrete-time system is its transfer function $H(z)$. If the ROC of $H(z)$ includes the unit circle, as it does in the case of a stable system, then $H(z)$ evaluated on the unit circle, that is, for $z = e^{j\omega}$, is the frequency response $H(e^{j\omega})$ of the system. In this project you will study various properties of a causal stable LTI discrete-time system. You will be concerned here with the evaluation of the frequency response from the transfer function, computation of the group delay of the system, implementation of the difference equation representing the system, determination of the impulse response from the transfer function, development of the pole-zero plot from the transfer function, and investigation of the stability of the LTI system from the pole-zero plot.

Project 4.1 Transfer Function Analysis

The frequency response of an LTI discrete-time system characterized by a difference equation of the form of Eq. (2.11) is given by Eq. (4.9). The frequency response can thus be easily evaluated at a set of discrete frequency points using the command freqz. In fact, Program P3_1 can be used for this purpose, as was done earlier in Question Q3.41 in the previous exercise.

Questions:

Q4.1 Modify Program P3_1 to compute and plot the magnitude and phase spectra of a moving average filter of Eq. (2.13) for three different values of length M and for $0 \le \omega \le 2\pi$. Justify the type of symmetries exhibited by the magnitude and phase spectra. What type of filter does it represent? Can you now explain the results of Question Q2.1?

The modified Program P3_1 can also be used to compute the frequency response of an LTI discrete-time system from its transfer function description as in Eq. (4.12).

Questions:

Q4.2 Using the modified Program P3_1 compute and plot the frequency response of a causal LTI discrete-time system with a transfer function given by

$$H(z) = \frac{0.15\,(1 - z^{-2})}{1 - 0.5\,z^{-1} + 0.7\,z^{-2}},\tag{4.36}$$

for $0 \le \omega \le \pi$. What type of filter does it represent?

Q4.3 Repeat Question Q4.3 for the following transfer function:

$$G(z) = \frac{0.15\,(1 - z^{-2})}{0.7 - 0.5\,z^{-1} + z^{-2}}.\tag{4.37}$$

What is the difference between the two filters of Eqs. (4.36) and (4.37), respectively? Which one will you choose for filtering and why?

Some applications require that the group delay of the LTI discrete-time system be approximately constant in the frequency band of interest to preserve the waveform of the signal components in the band. The group delay of a transfer function can be readily computed using the function grpdelay.

Question:

Q4.4 Using MATLAB compute and plot the group delay of the causal LTI discrete-time system with a transfer function given by

$$H(z) = \frac{z^{-1} - 1.2\,z^{-2} + z^{-3}}{1 - 1.3\,z^{-1} + 1.04\,z^{-2} - 0.222\,z^{-3}},\tag{4.38}$$

for $0 \le \omega \le \pi$.

The function `impz` can be used to compute the beginning part of the impulse response of a causal LTI discrete-time system. To this end, the program you wrote in answering Question Q3.50 can be employed again.

Question:

Q4.5   Using the program developed in Question Q3.50, compute and plot the first 100 samples of the impulse responses of the two filters of Eqs. (4.36) and (4.37), respectively. Comment on your results.

The pole-zero plot of a transfer function also provides insight into the behavior of an LTI discrete-time system. Such a plot can be readily obtained using the command `zplane`.

Question:

Q4.6   Using `zplane` develop the pole-zero plots of the two filters of Eqs. (4.36) and (4.37), respectively. Comment on your results.

4.5   Types of Transfer Functions

One of the main applications of digital signal processing is in filtering discrete-time signals to remove undesirable components. The frequency responses of the four types of ideal filters are shown in Figure 4.1. These filters have doubly infinite impulse responses and are not realizable. In many applications, fairly simple realizable approximations to these filters are quite adequate. In this project you will investigate the properties of some approximations to the ideal filters. In Laboratory Exercise 7 you will design filters to meet prescribed specifications.

Project 4.2   Filters

The impulse response $h_{LP}[n]$ of the ideal lowpass filter of Figure 4.1 given by Eq. (4.14) is doubly infinite and cannot be implemented. Hence a simple approximation is achieved by just truncating the impulse response to a finite number of terms. However, the truncated impulse response represents a noncausal filter. A causal approximation is then obtained by shifting the truncated filter impulse response to the right by $N/2$ samples, resulting in

$$\hat{h}_{LP}[n] = \frac{\sin \omega_c(n - N/2)}{\pi(n - N/2)}, \qquad 0 \le n \le N. \qquad (4.39)$$

The length of the filter is $N + 1$.

The following program, which uses the function `sinc`, can be used to compute the above approximation.

```
% Program P4_1
% Impulse Response of Truncated Ideal Lowpass Filter
```

```
clf;
fc = 0.25;
n = [-6.5:1:6.5];
y = 2*fc*sinc(2*fc*n);k = n+6.5;
stem(k,y);title('N = 13');axis([0 13 -0.2 0.6]);
xlabel('Time index n');ylabel('Amplitude'); grid
```

Questions:

Q4.7 Compute and plot the impulse response of the approximation to the ideal lowpass filter using Program P4_1. What is the length of the FIR lowpass filter? Which statement in Program P4_1 determines the filter length? Which parameter controls the cutoff frequency?

Q4.8 Modify Program P4_1 to compute and plot the impulse response of the FIR lowpass filter of Eq. (4.39) with a length of 20 and an angular cutoff frequency of $\omega_c = 0.45$.

Q4.9 Modify Program P4_1 to compute and plot the impulse response of the FIR lowpass filter of Eq. (4.39) with a length of 15 and an angular cutoff frequency of $\omega_c = 0.65$.

Q4.10 Write a MATLAB program to compute and plot the amplitude response of the FIR lowpass filter of Eq. (4.39). Using this program, plot the amplitude response for several values of N and comment on your results.

The moving-average filter of Eq. (2.13) also has a lowpass magnitude response as seen from the results of Question Q4.1. The simplest such filter is of length 2 and has a transfer function

$$H_0(z) = \frac{1}{2}(1 + z^{-1}).$$

(4.40)

It can be shown that this filter has a 3-dB cutoff frequency $\omega_c = \pi/2$. By cascading a number of these simple FIR lowpass filters, a lowpass filter with a sharper magnitude response can be obtained. A cascade of K sections of $H_0(z)$ has a 3-dB cutoff frequency at

$$\omega_c = 2 \cos^{-1}(2^{-1/2K}).$$

(4.41)

A slight modification of the difference equation of Eq. (2.13) yields a highpass filter whose transfer function is given by

$$H_1(z) = \frac{1}{M} \sum_{n=0}^{M-1} (-1)^n z^{-n}.$$

(4.42)

The function gain given below computes and plots the gain response in dB of a rational transfer function.

```
function [g,w] = gain(num,den)
% Computes the gain function in dB of a
% transfer function at 256 equally spaced points
% on the top half of the unit circle
% Numerator coefficients are in vector num
```

```
% Denominator coefficients are in vector den
% Frequency values are returned in vector w
% Gain values are returned in vector g
w = 0:pi/255:pi;
h = freqz(num,den,w);
g = 20*log10(abs(h));
```

Program P4_2 illustrates the use of the function gain in computing and plotting the gain response of a moving average lowpass filter.

```
% Program P4_2
% Gain Response of a Moving-Average Lowpass Filter
clf;
M = 2;
num = ones(1,M)/M;
[g,w] = gain(num,1);
plot(w/pi,g);grid
axis([0 1 -50 0.5])
xlabel('\omega /\pi');ylabel('Gain in dB');
title(['M = ', numstr(M)])
```

Questions:

Q4.11   Run Program P4_2 to compute and plot the gain response of a length-2 moving-average filter. From the plot verify that the 3-dB cutoff frequency is at $\pi/2$.

Q4.12   Modify Program P4_2 to compute and plot the gain response of a cascade of K length-2 moving-average filters. Using the modified program plot the gain response for a cascade of 3 sections and verify that the 3-dB cutoff frequency of the cascade is as given by Eq. (4.41).

Q4.13   Modify Program P4_2 to compute and plot the gain response of the highpass filter of Eq. (4.42). Run the modified program to plot the gain response for $M = 5$ and determine its 3-dB cutoff frequency from the plot.

In many applications, the simple first-order and second-order IIR filters described in R4.18 through R4.21 are adequate. If necessary, any of these filters can be cascaded to provide sharper gain response. Each of these filters has additional attractive properties to be demonstrated next. It can be shown easily that $H_{LP}(z)$ and $H_{HP}(z)$ of Eqs. (4.15) and (4.17), respectively, are allpass-complementary and power-complementary. The 3-dB cutoff frequency ω_c of both filters can be adjusted easily by changing the value of the multiplier coefficient α. Likewise, it can be shown easily that $H_{BP}(z)$ and $H_{BS}(z)$ of Eqs. (4.18) and (4.21), respectively, are allpass complementary and power-complementary. Here also the center and the notch frequencies ω_o of the two filters can be adjusted readily by changing the value of the multiplier coefficient β, and their 3-dB bandwidths can be changed by adjusting the value of the multiplier coefficient α.

Questions:

Q4.14   Design a first-order IIR lowpass and a first-order IIR highpass filter with a 3-dB angular cutoff frequency ω_c at 0.45π. Using MATLAB compute and plot their gain responses, and verify that the designed filters meet the specification. Using MATLAB show that the two filters are allpass-complementary and power-complementary.

Q4.15   Design an IIR lowpass filter with a 3-dB cutoff frequency ω_c at 0.3π by cascading 10 sections of the first-order IIR lowpass filter of Eq. (4.15). Compare its gain response with that of a first-order IIR lowpass filter designed for the same cutoff frequency.

Q4.16   Design a second-order IIR bandpass filter with a center frequency ω_o at 0.61π and a 3-dB bandwidth of 0.15π. Since Eq. (4.20) is a quadratic equation in α, there will be two values of the parameter α yielding the same value of the 3-dB bandwidth, resulting in two different expressions for the transfer function $H_{BP}(z)$. Using the function zplane, develop the pole-zero plots of the two designs obtained and choose the design that results in a stable transfer function. Using MATLAB compute and plot its gain response, and verify that the filter you designed indeed meets the given specifications. Now using the values of the parameters α and β of the stable IIR bandpass transfer function designed, develop the expression for a second-order IIR bandstop transfer function $H_{BS}(z)$. Using MATLAB show that $H_{BP}(z)$ and $H_{BS}(z)$ are both allpass-complementary and power-complementary.

If $H(z)$ is the transfer function of an FIR or IIR digital filter, the filter obtained by replacing each delay in the realization of $H(z)$ by L delays has a transfer function $G(z) = H(z^L)$. Thus, the new filter has a frequency response that is a periodic function of ω with a period $2\pi/L$. Such filters are generally called *comb filters* and find applications in rejecting periodic interferences.

Questions:

Q4.17   Using MATLAB compute and plot the magnitude response of a comb filter obtained from a prototype FIR lowpass filter of Eq. (4.38) for different values of L. Show that the new filter has multiple notches at $\omega = \omega_k = (2k + 1)\pi/L$ and has L peaks in its magnitude response at $\omega = \omega_k = 2k\pi/L, k = 0, 1, \ldots, L - 1$.

Q4.18   Using MATLAB compute and plot the magnitude response of a comb filter obtained from a prototype FIR highpass filter of Eq. (4.41) with $M = 2$ and for different values of L. Determine the locations of the notches and the peaks of the magnitude response of this type of comb filter.

Many applications require the use of digital filters with either linear phase or zero phase. Zero-phase filtering cannot be implemented using a causal digital filter. However the noncausal implementation indicated in Figure 4.2 can be employed for zero-phase filtering using either an FIR or an IIR digital filter. The MATLAB M-file filtfilt implements this type of zero-phase filtering scheme. It has also been designed to minimize the start-up transients. Since it uses the same filter for both the forward and reverse directions, the order of the filter implemented is twice that of the basic filter used and should be taken into account in any application of this function. However, FIR filters with an exact linear phase property can be designed and will be discussed in Laboratory Exercise 7. Four types of linear phase FIR filters are defined (see R4.23). Program P4_3 can be used to investigate the properties of these filters. It first generates the plots of the impulse response sequence of each of the four types, then generates the pole-zero plots, and finally displays the zero locations.

```
% Program P4_3
% Zero Locations of Linear Phase FIR Filters
clf;
b = [1 -8.5 30.5 -63];
num1 = [b 81 fliplr(b)];
num2 = [b 81 81 fliplr(b)];
num3 = [b 0 -fliplr(b)];
num4 = [b 81 -81 -fliplr(b)];
n1 = 0:length(num1)-1;
n2 = 0:length(num2)-1;
subplot(2,2,1); stem(n1,num1);
xlabel('Time index n');ylabel('Amplitude'); grid;
title('Type 1 FIR Filter');
subplot(2,2,2); stem(n2,num2);
xlabel('Time index n');ylabel('Amplitude'); grid;
title('Type 2 FIR Filter');
subplot(2,2,3); stem(n1,num3);
xlabel('Time index n');ylabel('Amplitude'); grid;
title('Type 3 FIR Filter');
subplot(2,2,4); stem(n2,num4);
xlabel('Time index n');ylabel('Amplitude'); grid;
title('Type 4 FIR Filter');
pause
subplot(2,2,1); zplane(num1,1);
title('Type 1 FIR Filter');
subplot(2,2,2); zplane(num2,1);
title('Type 2 FIR Filter');
subplot(2,2,3); zplane(num3,1);
title('Type 3 FIR Filter');
subplot(2,2,4); zplane(num4,1);
title('Type 4 FIR Filter');
disp('Zeros of Type 1 FIR Filter are');
disp(roots(num1));
disp('Zeros of Type 2 FIR Filter are');
disp(roots(num2));
disp('Zeros of Type 3 FIR Filter are');
disp(roots(num3));
disp('Zeros of Type 4 FIR Filter are');
disp(roots(num4));
```

Questions:

Q4.19 Run Program P4_3 and generate the plots of the impulse response sequence of each type of linear-phase FIR filter. What are the lengths of each FIR filter? Verify the symmetry properties of the impulse response sequences. Verify next the zero locations of these filters. Using MATLAB compute and plot the phase response of each of these filters and verify the linear-phase property of each. What are the group delays of these filters?

Q4.20 Replace the vector b in Program P4_3 with b = [1.5 -3.25 5.25 -4] and repeat Question Q4.19.

As we shall demonstrate in Chapter 9, the bounded-real property of a causal transfer function is key to its low passband sensitivity realization. To test the bounded-real property, you must test the stability of the transfer function first and then determine the maximum value of its magnitude response. Note any causal stable transfer function can be converted into a bounded-real function by simple scaling.

Questions:

Q4.21 Using MATLAB determine whether the following transfer function is bounded-real:

$$H_1(z) = \frac{1.5(1 + z^{-1})}{1 + 5\,z^{-1} + 6\,z^{-2}}. \tag{4.43}$$

If it is not a bounded-real function, determine another transfer function $H_2(z)$ that is bounded-real and has the same magnitude as $H_1(z)$.

Q4.22 Using MATLAB determine whether the following transfer function is bounded-real:

$$G_1(z) = \frac{1.5(1 - z^{-1})}{2 + z^{-1} + z^{-2}}. \tag{4.44}$$

If it is not a bounded-real function, determine another transfer function $G_2(z)$ that is bounded-real and has the same magnitude as $G_1(z)$.

4.6 Stability Test

The stability of an IIR causal digital filter is an important design requirement. A causal IIR filter is stable if all poles of its transfer function are inside the unit circle. The MATLAB function zplane can be used to check the pole locations of an IIR transfer function. However, if one or more poles are very close to the unit circle or on the unit circle, the pole-zero plot is not sufficient to test the stability of the corresponding transfer function. A more accurate stability test is based on the algorithm given in R4.32. Program P4_4 implements this using the function poly2rc.

```
% Program P4_4
% Stability Test
clf;·
```

```
den = input('Denominator coefficients = ');
ki = poly2rc(den);
disp('Stability test parameters are');
disp(ki);
```

Questions:

Q4.23   Using MATLAB generate the pole-zero plots of the following two causal transfer functions.

$$H_1(z) = \frac{1}{1 - 1.848\,z^{-1} + 0.85\,z^{-2}},$$

$$H_2(z) = \frac{1}{1 - 1.851\,z^{-1} + 0.85\,z^{-2}}.$$

Can you infer their stability by examining the generated pole-zero plots?

Q4.24   Test the stability of the two transfer functions in Question Q4.23 using Program P4_4. Which one of the two transfer functions is stable?

Q4.25   Using Program P4_4 determine whether or not all the roots of the following polynomial are inside the unit circle:

$$D(z) = 1 + 2.5\,z^{-1} + 2.5\,z^{-2} + 1.25\,z^{-3} + 0.3125\,z^{-4} + 0.03125\,z^{-5}.$$

Q4.26   Using Program P4_4 determine whether or not all the roots of the following polynomial are inside the unit circle:

$$D(z) = 1 + 0.2\,z^{-1} + 0.3\,z^{-2} + 0.4\,z^{-3} + 0.5\,z^{-4} + 0.6\,z^{-5}.$$

4.7   Background Reading

[1] A. Antoniou. *Digital Filters: Analysis, Design, and Applications.* McGraw-Hill, New York NY, second edition, 1993. Ch. 3.

[2] E. Cunningham. *Digital Filtering: An Introduction.* Houghton-Mifflin, Boston MA, 1992. Ch. 8.

[3] D.J. DeFatta, J.G. Lucas, and W.S. Hodgkiss. *Digital Signal Processing: A System Design Approach.* Wiley, New York NY, 1988. Ch. 9.

[4] L.B. Jackson. *Digital Filters and Signal Processing.* Kluwer, Boston MA, third edition, 1996. Ch. 11.

[5] R. Kuc. *Introduction to Digital Signal Processing.* McGraw-Hill, New York NY, 1988. Ch. 10.

[6] L.C. Ludeman. *Fundamentals of Digital Signal Processing.* Harper & Row, New York NY, 1986. Secs. 1.3.5, 2.3, 2.4.

[7] S.K. Mitra. *Digital Signal Processing: A Computer-Based Approach*. McGraw-Hill, New York NY, 1998. Sec. 8.4 and Ch. 9.

[8] A.V. Oppenheim and R.W. Schafer. *Discrete-Time Signal Processing*. Prentice-Hall, Englewood Cliffs NJ, 1989. Secs. 6.7–6.10.

[9] S.J. Orfanidis. *Introduction to Signal Processing*. Prentice-Hall, Englewood Cliffs NJ, 1996. Ch. 2, Sec. 7.6.

[10] B. Porat. *A Course in Digital Signal Procesing*. Wiley, New York NY, 1996. Secs. 2.7, 7.6, 8.1, 8.2, 8.4, 9.1.

[11] J.G. Proakis and D.G. Manolakis. *Digital Signal Processing: Principles, Algorithms, and Applications*. Prentice-Hall, Englewood Cliffs NJ, third edition, 1996. Secs. 1.4, 7.5–7.7.

[12] R.A. Roberts and C.T. Mullis. *Digital Signal Processing*. Addison-Wesley, Reading MA, 1987. Ch. 9.

Digital Processing of Continuous-Time Signals 5

5.1 Introduction

Digital signal processing algorithms are often used to process continuous-time signals. To this end, it is necessary to convert a continuous-time signal into an equivalent discrete-time signal, apply the necessary digital signal processing algorithm to it, and then convert back the processed discrete-time signal into an equivalent continuous-time signal. In the ideal case, the conversion of a continuous-time signal into a discrete-time form is implemented by periodic sampling, and to prevent *aliasing*, an analog *anti-aliasing filter* is often placed before sampling to bandlimit the continuous-time signal. The conversion of a discrete-time signal into a continuous-time signal requires an analog *reconstruction filter* . In this exercise you will investigate the effect of sampling in the time-domain and the frequency-domain, respectively, and the design of the analog filters. In addition, you will learn the basics of analog-to-digital and digital-to-analog conversions.

5.2 Background Review

R5.1 Let $g_a(t)$ be a continuous-time signal that is sampled uniformly at $t = nT$ generating the sequence $g[n]$ where

$$g[n] = g_a(nT), \qquad -\infty < n < \infty, \qquad (5.1)$$

with T being the *sampling period* . The reciprocal of T is called the *sampling frequency* F_T, that is, $F_T = 1/T$. Now, the frequency-domain representation of $g_a(t)$ is given by its continuous-time Fourier transform $G_a(j\Omega)$,

$$G_a(j\Omega) = \int_{-\infty}^{\infty} g_a(t)\, e^{-j\Omega t}\, dt, \qquad (5.2)$$

whereas the frequency-domain representation of $g[n]$ is given by its discrete-time Fourier transform $G(e^{j\omega})$,

$$G(e^{j\omega}) = \sum_{n=-\infty}^{\infty} g[n]\, e^{-j\omega n}. \qquad (5.3)$$

The relation between $G_a(j\Omega)$ and $G(e^{j\omega})$ is given by

$$G(e^{j\omega}) = \frac{1}{T} \sum_{k=-\infty}^{\infty} G_a(j\Omega - jk\Omega_T)|_{\Omega=\omega/T}$$

71

$$= \frac{1}{T} \sum_{k=-\infty}^{\infty} G_a(j\frac{\omega}{T} - jk\Omega_T) = \frac{1}{T} \sum_{k=-\infty}^{\infty} G_a(j\frac{\omega}{T} - j\frac{2\pi k}{T}), \qquad (5.4)$$

which can be expressed alternately as

$$G(e^{j\Omega T}) = \frac{1}{T} \sum_{k=-\infty}^{\infty} G_a(j\Omega - jk\Omega_T). \qquad (5.5)$$

R5.2   *Sampling Theorem* - Let $g_a(t)$ be a bandlimited signal with $G_a(j\Omega) = 0$ for $|\Omega| > \Omega_m$. Then $g_a(t)$ is uniquely determined by its samples $g_a(nT), n = 0, 1, 2, 3, \ldots$, if

$$\Omega_T > 2\Omega_m, \qquad (5.6)$$

where

$$\Omega_T = \frac{2\pi}{T}. \qquad (5.7)$$

Given $\{g[n]\} = \{g_a(nT)\}$, we can recover $g_a(t)$ exactly by generating an impulse train $g_p(t)$ of the form

$$g_p(t) = g_a(t)\, p(t) = \sum_{n=-\infty}^{\infty} g_a(nT)\delta(t - nT), \qquad (5.8)$$

and then passing $g_p(t)$ through an ideal lowpass filter $H_r(j\Omega)$ with a gain T and a cutoff frequency Ω_c greater than Ω_m and less than $\Omega_T - \Omega_m$, that is,

$$\Omega_m < \Omega_c < (\Omega_T - \Omega_m). \qquad (5.9)$$

The highest frequency Ω_m contained in $g_a(t)$ is usually called the *Nyquist frequency* as it determines the minimum sampling frequency $\Omega_T > 2\Omega_m$ that must be used to fully recover $g_a(t)$ from its sampled version. The frequency $2\Omega_m$ is called the *Nyquist rate*.

If the sampling rate is higher than the Nyquist rate, it is called *oversampling* . On the other hand, if the sampling rate is lower than the Nyquist rate, it is called *undersampling*. Finally, if the sampling rate is exactly equal to the Nyquist rate, it is called *critical sampling*.

R5.3   Now, the impulse response $h_r(t)$ of an ideal analog lowpass filter is simply obtained by taking the inverse Fourier transform of its frequency response $H_r(j\Omega)$:

$$H_r(j\Omega) = \begin{cases} T, & |\Omega| \le \Omega_c, \\ 0, & |\Omega| > \Omega_c, \end{cases} \qquad (5.10)$$

and is given by

$$h_r(t) = \frac{1}{2\pi} \int_{-\infty}^{\infty} H_r(j\Omega)\, e^{j\Omega t}\, dt$$

$$= \frac{T}{2\pi} \int_{-\Omega_c}^{\Omega_c} e^{j\Omega t} d\Omega = \frac{\sin(\Omega_c t)}{\Omega_T t/2}, \qquad -\infty \le t \le \infty. \qquad (5.11)$$

Now the impulse train $g_p(t)$ is given by

$$g_p(t) = \sum_{n=-\infty}^{\infty} g[n]\, \delta(t - nT). \tag{5.12}$$

Therefore, the output $\hat{g}_a(t)$ of the ideal lowpass filter is given by the convolution of $g_p(t)$ with the impulse response $h_r(t)$ of the analog reconstruction filter :

$$\hat{g}_a(t) = \sum_{n=-\infty}^{\infty} g[n]\, h_r(t - nT). \tag{5.13}$$

Substituting $h_r(t)$ from Eq. (5.11) in Eq. (5.13) and assuming for simplicity $\Omega_c = \Omega_T/2 = \pi/T$, we arrive at

$$\hat{g}_a(t) = \sum_{n=-\infty}^{\infty} g[n]\, \frac{\sin[\pi(t - nT)/T]}{\pi(t - nT)/T}. \tag{5.14}$$

R5.4   The filter specifications are usually stated in terms of its magnitude response. For example, the magnitude $|H_a(j\Omega)|$ of an analog lowpass filter is usually specified as indicated in Figure 5.1. In the *passband* defined by $0 \leq \Omega \leq \Omega_p$, we require

$$1 - \delta_p \leq |H_a(j\Omega)| \leq 1 + \delta_p, \quad \text{for} \quad |\Omega| \leq \Omega_p, \tag{5.15}$$

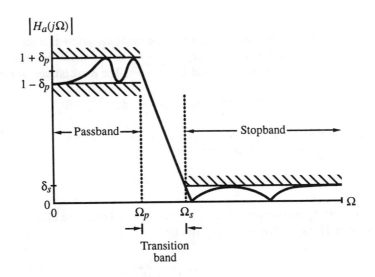

Figure 5.1   Typical magnitude response specifications for an analog lowpass filter.

or, in other words, the magnitude approximates unity within an error of $\pm\delta_p$. In the *stopband* , defined by $\Omega_s \leq |\Omega| \leq \infty$, we require

$$|H_a(j\Omega)| \leq \delta_s, \quad \text{for} \quad \Omega_s \leq |\Omega| \leq \infty, \tag{5.16}$$

implying that the magnitude approximate zero within an error of δ_s. The frequencies Ω_p and Ω_s are called, respectively, the *passband edge frequency* and the *stopband edge frequency* . The limits of the tolerances in the passband and stopband, δ_p and δ_s, are called *ripples* .

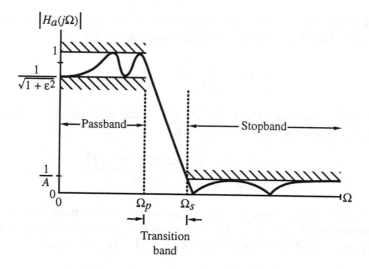

Figure 5.2 Normalized magnitude response specifications for an analog lowpass filter.

R5.5 In most applications, the analog filter specifications are given as indicated in Figure 5.2. Here, in the *passband* defined by $0 \leq \Omega \leq \Omega_p$, the maximum and the minimum values of the magnitude are, respectively, unity and $1/\sqrt{1 + \epsilon^2}$. The *peak passband ripple* is

$$R_p = 20 \log_{10} \sqrt{1 + \epsilon^2} \ \text{dB}. \tag{5.17}$$

The maximum stopband ripple in the *stopband*, defined by $\Omega_s \leq \Omega \leq \infty$, is denoted by $1/A$. The *minimum stopband attenuation* is therefore given by

$$R_s = 20 \log_{10} A \ \text{dB}. \tag{5.18}$$

R5.6 *Butterworth Approximation.* The magnitude-squared response of an analog lowpass Butterworth filter $H_a(s)$ of Nth order is given by

$$|H_a(j\Omega))|^2 = \frac{1}{1 + (\frac{\Omega}{\Omega_c})^{2N}}. \tag{5.19}$$

The Butterworth filter is said to have a maximally flat magnitude at $\Omega = 0$ as the first $2N - 1$ derivatives of $|H_a(j\Omega))|^2$ at $\Omega = 0$ are equal to zero. At $\Omega = \Omega_c$, the gain $\mathcal{G}(\Omega) = 10 \log_{10} |H_a(j\Omega)|^2$ is 3 dB less than that at $\Omega = 0$ and, hence, Ω_c is called the *3-dB cutoff frequency*.

The two parameters completely characterizing a Butterworth lowpass filter are therefore the 3-dB cutoff frequency Ω_c and the order N. These are determined from the specified *passband edge* Ω_p, the *stopband edge* Ω_s, the *peak passband ripple* R_p in dB, and the *minimum stopband attenuation* R_s in dB.

The transfer function of the Butterworth lowpass filter is of the form

$$H_a(s) = \frac{K}{\sum_{\ell=0}^{N} a_\ell s^\ell} = \frac{K}{\prod_{\ell=1}^{N}(s - p_\ell)}. \qquad (5.20)$$

R5.7 *Type 1 Chebyshev Approximation.* The Type 1 Chebyshev lowpass transfer function $H_a(s)$ has a magnitude response given by

$$|H_a(j\Omega)|^2 = \frac{1}{1 + \varepsilon^2 T_N^2(\omega - \Omega_p)}, \qquad (5.21)$$

where $T_N(\Omega)$ is the Chebyshev polynomial of order N:

$$T_N(\Omega) = \begin{cases} \cos(N \cos^{-1}\Omega), & |\Omega| \le 1, \\ \cosh(N \cosh^{-1}\Omega), & |\Omega| > 1. \end{cases} \qquad (5.22)$$

The above polynomial can also be derived via a recurrence relation given by

$$T_r(\Omega) = 2\Omega\, T_{r-1}(\Omega) - T_{r-2}(\Omega), \qquad r \ge 2, \qquad (5.23)$$

with $T_0(\Omega) = 1$ and $T_1(\Omega) = \Omega$.

The order N of the Type 1 Chebyshev lowpass filter is determined from the specified passband edge Ω_p, the stopband edge Ω_s, the peak passband ripple R_p in dB, and the minimum stopband attenuation R_s in dB. The transfer function $H_a(s)$ is again of the form of Eq. (5.20).

R5.8 *Type 2 Chebyshev Approximation.* The square-magnitude response expression here is given by

$$|H_a(j\Omega))|^2 = \frac{1}{1 + \varepsilon^2 \left[\frac{T_n(\Omega_s/\Omega_p)}{T_n(\Omega_s/\Omega)}\right]^2}. \qquad (5.24)$$

The transfer function of a Type 2 Chebyshev lowpass filter is no longer an all-pole function, as it has both poles and zeros. It is of the form

$$H_a(s) = K\frac{\sum_{\ell=0}^{N} b_\ell s^\ell}{\sum_{\ell=0}^{N} a_\ell s^\ell} = K\frac{\prod_{\ell=1}^{N}(s - z_\ell)}{\prod_{\ell=1}^{N}(s - p_\ell)}. \qquad (5.25)$$

The zeros z_ℓ here are on the $j\Omega$-axis. The order N of the Type 2 Chebyshev lowpass filter is determined from the specified passband edge Ω_p, the stopband edge Ω_s, the peak passband ripple R_p in dB, and the minimum stopband attenuation R_s in dB.

R5.9 *Elliptic Approximation.* The square-magnitude response of an elliptic lowpass filter is given by

$$|H_a(j\Omega))|^2 = \frac{1}{1 + \varepsilon^2 R_N^2(\Omega/\Omega_p)}, \tag{5.26}$$

where $R_N(\Omega)$ is a rational function of order N satisfying the property $R_N(1/\Omega) = 1/R_N(\Omega)$ with the roots of its numerator lying within the interval $0 < \Omega < 1$ and the roots of its denominator lying in the interval $1 < \Omega < \infty$. The order N of the elliptic lowpass filter is determined from the specified passband edge Ω_p, the stopband edge Ω_s, the peak passband ripple R_p in dB, and the minimum stopband attenuation R_s in dB.

R5.10 *A Comparison of Filter Types.* The performances of the above types of approximations are compared next by examining the frequency responses of the normalized Butterworth, Chebyshev, and elliptic analog lowpass filters of the same order. The passband ripples of the Type 1 Chebyshev and the equiripple filters are assumed to be the same, while the minimum stopband attenuations of the Type 2 Chebyshev and the equiripple filters are assumed to be the same. The filter specifications used for comparison are as follows: filter order of 6, passband edge at $\Omega = 1$, maximum passband deviation of 1 dB, and minimum stopband attenuation of 40 dB. The frequency responses computed using MATLAB are plotted in Figure 5.3.

As can be seen from Figure 5.3, the Butterworth filter has the widest transition band with a monotonically decreasing gain response. Both types of Chebyshev filters have a transition band of equal width that is smaller than that of the Butterworth filter but greater than that of the elliptic filter. The Type 1 Chebyshev filter provides a slightly faster roll-off in the transition band than the Type 2 Chebyshev filter. The magnitude response of the Type 2 Chebyshev filter in the passband is nearly identical to that of the Butterworth filter. The elliptic filter has the narrowest transition band with an equiripple passband and an equiripple stopband response.

The Butterworth and the Chebyshev filters have a nearly linear phase response over about three-fourths of the passband, whereas the elliptic filter has a nearly linear phase response over about half of the passband.

R5.11 *Binary Number Representation.* In a binary representation, a number is represented using the symbols 0 and 1, called *bits*. The binary point separates the integer part from the fractional part. An additional *sign bit* is placed to the left of the integer part to indicate the sign of the number. For a positive number the sign bit is a 0 and for a negative number it is a 1. Three commonly used forms of binary number representations are the *sign-magnitude*, *ones'-complement* and *two's-complement forms*.

A binary number consisting of I integer bits, F fractional bits, and a sign bit is of the form $s\, a_{I-1} a_{I-2} \dots a_1 a_0 \triangle a_{-1} a_{-2} \dots a_{-F}$ where $\triangle$ denotes the binary point, and each bit a_k and the sign bit s are either a 0 or a 1. The bit a_{I-1} is called the *most significant bit*, abbreviated as *MSB*, and the bit a_{-F} is called the *least significant bit*, abbreviated as *LSB*. In a fixed-point binary representation, the position of the binary point is always at the same place for all numbers . In digital signal processing applications, fixed-point numbers are

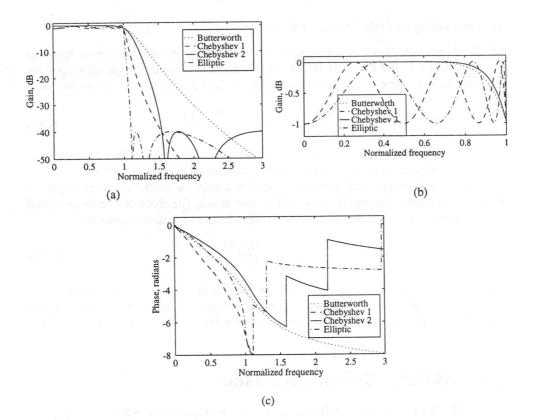

Figure 5.3 A comparison of the frequency responses of the four types of analog lowpass filter: (a) gain responses, (b) passband details, and (c) phase responses.

always represented as fractions.

The decimal equivalent of the binary fraction $s_\triangle a_{-1} a_{-2} \ldots a_{-F}$ in sign-magnitude form is a positive number with a magnitude given by $\sum_{k=1}^{F} a_{-k} 2^{-k}$ if $s = 0$ and is a negative number with a magnitude given by $\sum_{k=1}^{F} a_{-k} 2^{-k}$ if $s = 1$.

In the ones'-complement form, the decimal equivalent of a positive or a negative fraction $s_\triangle a_{-1} a_{-2} \ldots a_{-F}$ is given by $-s(1 - 2^{-F}) + \sum_{k=1}^{F} a_{-k} 2^{-k}$. In this form, a positive fraction is represented as in the sign-magnitude form while its negative is represented by complementing each bit of the binary representation (i.e., by replacing each 0 with a 1 and vice-versa) of the positive fraction in the sign-magnitude form.

In the two's-complement form, the decimal equivalent of a positive or a negative fraction $s_\triangle a_{-1} a_{-2} \ldots a_{-F}$ is given by $-s + \sum_{k=1}^{F} a_{-k} 2^{-k}$. In this form, the positive fraction is represented as in the sign-magnitude form while its negative is represented by complementing each bit of the binary representation of the positive fraction in the sign-magnitude form, and adding a 1 to the LSB, the Fth bit.

R5.12   A practical D/A converter first develops an analog periodic pulse-train from the digtal input signal and then converts it into a staircase-like analog waveform $y_z(t)$ by a zero-order hold circuit. The continuous-time Fourier transform of the zero-order hold circuit is given by

$$H_z(j\Omega) = e^{-j\frac{\Omega T}{2}} \left[\frac{sin(\Omega T/2)}{\Omega/2} \right], \tag{5.27}$$

where T is the sampling period of the digital signal. The zero-order hold circuit thus has a magnitude response with a lowpass characteristic with zeros at integer multiples of $\Omega = 1/T$ and introduces amplitude distortion, called *droop*, in the magnitude response of the discrete-time system between the A/D and D/A converters. The droop can be compensated by designing the analog reconstruction filter with a frequency response given by

$$\hat{H}_r(j\Omega) = \frac{H_r(j\Omega)}{H_z(j\Omega)}, \tag{5.28}$$

where $H_r(j\Omega)$ is the frequency response of the ideal analog reconstruction lowpass filter. Alternately, the droop can be compensated by including prior to D/A conversion a digital compensation filter with a magnitude response that is the inverse of that of the zero-order hold circuit.

5.3   MATLAB Commands Used

The MATLAB commands you will encounter in this exercise are as follows:

General Purpose Commands

     length   size

Operators and Special Characters

     :   .   +   –   *   /   ;
     %   ==   ~   & |

Elementary Matrices and Matrix Manipulation

     '   ones   linspace   pi

Elementary Functions

     abs   cos   exp

Two-Dimensional Graphics

     axis   plot   stem   title   xlabel
     ylabel

General Purpose Graphics Functions

clf	grid	plot	stem	subplot

Signal Processing Toolbox

butter	buttord	cheb1ord	cheb2ord
cheby1	cheby2	ellip	elliford
freqs	freqz	sinc	

For additional information on these commands, see the *MATLAB Reference Guide* [Mat94] and the *Signal Processing Toolbox User's Guide* [Mat96] or type help commandname in the Command window. A brief explanation of the MATLAB functions used here can be found in Appendix B.

5.4   The Sampling Process in the Time-Domain

The purpose of this section is to study the relation in the time-domain between a continuous-time signal $x_a(t)$ and the discrete-time signal $x[n]$ generated by a periodic sampling of $x_a(t)$.

Project 5.1   Sampling of a Sinusoidal Signal

In this project you will investigate the sampling of a continuous-time sinusoidal signal $x_a(t)$ at various sampling rates. Since MATLAB cannot strictly generate a continuous-time signal, you will generate a sequence $\{x_a(nT_H)\}$ from $x_a(t)$ by sampling it at a very high rate T_H such that the samples are very close to each other. A plot of $x_a(nT_H)$ using the plot command will then look like a continuous-time signal.

```
% Program P5_1
% Illustration of the Sampling Process
% in the Time-Domain
clf;
t = 0:0.0005:1;
f = 13;
xa = cos(2*pi*f*t);
subplot(2,1,1)
plot(t,xa);grid
xlabel('Time, msec');ylabel('Amplitude');
title('Continuous-time signal x_{a}(t)');
axis([0 1 -1.2 1.2])
subplot(2,1,2);
T = 0.1;
n = 0:T:1;
xs = cos(2*pi*f*n);
k = 0:length(n)-1;
```

```
stem(k,xs); grid
xlabel('Time index n');ylabel('Amplitude');
title('Discrete-time signal x[n]');
axis([0 (length(n)-1) -1.2 1.2])
```

Questions:

Q5.1   Run Program P5_1 to generate both the continuous-time signal and its sampled version, and display them.

Q5.2   What is the frequency in Hz of the sinusoidal signal? What is the sampling period in seconds?

Q5.3   Explain the effects of the two `axis` commands.

Q5.4   Run Program P5_1 for four other values of the sampling period with two lower and two higher than that listed in Program P5_1. Comment on your results.

Q5.5   Repeat Program P5_1 by changing the frequency of the sinusoidal signal to 3 Hz and 7 Hz, respectively. Is there any difference between the corresponding equivalent discrete-time signals and the one generated in Question Q5.1? If not, why not?

Project 5.2   Aliasing Effect in the Time-Domain

In this project you will generate a continuous-time equivalent $y_a(t)$ of the discrete-time signal $x[n]$ generated in Program P5_1 to investigate the relation between the frequency of the sinusoidal signal $x_a(t)$ and the sampling period. To generate the reconstructed signal $y_a(t)$ from $x[n]$, we pass $x[n]$ through an ideal lowpass filter that in turn can be implemented according to Eq. (5.11) (see R5.3). If Eq. (5.11) is computed at closely spaced values of t, a plot of $y_a(t)$ will resemble a continuous-time signal. In order to implement this equation on MATLAB, the summation in Eq. (5.11) needs to be replaced with a finite sum, and hence we can generate only an approximation to the desired reconstructed continuous-time signal $y_a(t)$.

```
% Program P5_2
% Illustration of Aliasing Effect in the Time-Domain
% Program adapted from [Kra94] with permission from
% The Mathworks, Inc., Natick, MA.
clf;
T = 0.1;f = 13;
n = (0:T:1)';
xs = cos(2*pi*f*n);
t = linspace(-0.5,1.5,500)';
ya = sinc((1/T)*t(:,ones(size(n))) - (1/T)*n(:,ones(size(t)))')*xs;
plot(n,xs,'o',t,ya);grid;
xlabel('Time, msec');ylabel('Amplitude');
title('Reconstructed continuous-time signal y_{a}(t)');
axis([0 1 -1.2 1.2]);
```

Questions:

Q5.6 Run Program P5_2 to generate both the discrete-time signal $x[n]$ and its continuous-time equivalent $y_a(t)$, and display them.

Q5.7 What is the range of t and the value of the time increment in Program P5_2 ? What is the range of t in the plot? Change the range of t so as to display the full range $y_a(t)$ being computed in the above program and run Program P5_2 again. Comment on the plot generated after this change.

Q5.8 Restore the original display range and repeat Program P5_2 by changing the frequency of the sinusoidal signal to 3 Hz and 7 Hz, respectively. Is there any difference between the corresponding equivalent discrete-time signals and the one generated in Question Q5.6? If not, why not?

5.5 Effect of Sampling in the Frequency-Domain

Project 5.3 Aliasing Effect in the Frequency-Domain

The relation between the continuous-time Fourier transform (CTFT) of an arbitrary band-limited continuous-time signal and the discrete-time Fourier transform (DTFT) of the discrete-time signal is investigated next in this project. In order to convert a continuous-time signal $x_a(t)$ into an equivalent discrete-time signal $x[n]$, the former must be band-limited in the frequency-domain (see R5.2). To illustrate the effect of sampling in the frequency-domain we choose an exponentially decaying continuous-time signal with a CTFT that is approximately band-limited.

```
% Program P5_3
% Illustration of the Aliasing Effect
% in the Frequency-Domain
clf;
t = 0:0.005:10;
xa = 2*t.*exp(-t);
subplot(2,2,1)
plot(t,xa);grid
xlabel('Time, msec');ylabel('Amplitude');
title('Continuous-time signal x_{a}(t)');
subplot(2,2,2)
wa = 0:10/511:10;
ha = freqs(2,[1 2 1],wa);
plot(wa/(2*pi),abs(ha));grid;
xlabel('Frequency, kHz');ylabel('Amplitude');
title('|X_{a}(j\Omega)|');
axis([0 5/pi 0 2]);
subplot(2,2,3)
T = 1;
n = 0:T:10;
```

```
xs = 2*n.*exp(-n);
k = 0:length(n)-1;
stem(k,xs);grid;
xlabel('Time index n');ylabel('Amplitude');
title('Discrete-time signal x[n]');
subplot(2,2,4)
wd = 0:pi/255:pi;
hd = freqz(xs,1,wd);
plot(wd/(T*pi), T*abs(hd));grid;
xlabel('Frequency, kHz');ylabel('Amplitude');
title('|X(e^{j\omega})|');
axis([0 1/T 0 2])
```

Questions:

Q5.9   What is the continuous-time function $x_a(t)$ in Program P5_3? How is the CTFT of $x_a(t)$ being computed?

Q5.10   Run Program P5_3 to generate and display both the discrete-time signal and its continuous-time equivalent, and their respective Fourier transforms. Is there any visible effect of aliasing?

Q5.11   Repeat Program P5_3 by increasing the sampling period to 1.5. Is there any visible effect of aliasing?

Q5.12   Modify Program P5_3 for the case of $x_a(t) = e^{-\pi t^2}$ and repeat Questions Q5.10 and Q5.11.

5.6   Analog Lowpass Filters

Analog lowpass filters are employed as anti-aliasing filters and as anti-imaging filters in the digital processing of continuous-time signals. In this section you will learn the design of the four types of analog lowpass filters summarized in R5.6 through R5.9.

Project 5.4   Design of Analog Lowpass Filters

The first step in the design of any of these filters is the determination of the filter order N and the appropriate cutoff frequency Ω_c . These parameters can be determined using the MATLAB commands `buttord` for the Butterworth filter, `cheb1ord` for the Type 1 Chebyshev filter, `cheb2ord` for the Type 2 Chebyshev filter, and `ellipord` for the elliptic filter. Ω_c is the 3-dB cutoff frequency for the Butterworth filter, the passband edge for the Type 1 Chebyshev filter, the stopband edge for the Type 2 Chebyshev filter, and the passband edge for the elliptic filter. For the design of filters MATLAB commands are `butter` for the Butterworth filter, `cheby1` for the Type 1 Chebyshev filter, `cheby2` for the Type 2 Chebyshev filter, and `ellip` for the elliptic filter.

Program P5_4 can be used for the design of the Butterworth lowpass filter.

```
% Program P5_4
% Design of Analog Lowpass Filter
clf;
Fp = 3500;Fs = 4500;
Wp = 2*pi*Fp; Ws = 2*pi*Fs;
[N, Wn] = buttord(Wp, Ws, 0.5, 30,'s');
[b,a] = butter(N, Wn, 's');
wa = 0:(3*Ws)/511:3*Ws;
h = freqs(b,a,wa);
plot(wa/(2*pi), 20*log10(abs(h)));grid
xlabel('Frequency, Hz');ylabel('Gain, dB');
title('Gain response');
axis([0 3*Fs -60 5]);
```

Questions:

Q5.13   What are the passband ripple Rp in dB and the minimum stopband attenuation Rs in dB in Program P5_4? What are the passband and the stopband edge frequencies in Hz?

Q5.14   Run Program P5_4 and display the gain response. Does the filter as designed meet the given specifications? What are the filter order N and the 3-dB cutoff frequency in Hz of the filter as designed?

Q5.15   Using cheb1ord and cheby1 modify Program P5_4 to design a Type 1 Chebyshev lowpass filter meeting the same specifications as in Program P5_4 . Run the modified program and display the gain response. Does the filter as designed meet the given specifications? What are the filter order N and the passband edge frequency in Hz of the filter as designed?

Q5.16   Using cheb2ord and cheby2 modify Program P5_4 to design a Type 2 Chebyshev lowpass filter meeting the same specifications as in Program P5_4. Run the modified program and display the gain response. Does the filter as designed meet the given specifications? What are the filter order N and the stopband edge frequency in Hz of the filter as designed?

Q5.17   Using ellipord and ellip modify Program P5_4 to design an elliptic lowpass filter meeting the same specifications as in Program P5_4. Run the modified program and display the gain response. Does the filter as designed meet the given specifications? What are the filter order N and the passband edge frequency in Hz of the filter as designed?

5.7   A/D and D/A Conversions

In this section you will learn the basics of analog-to-digital and digital-to-analog conversions, and binary representations of decimal numbers.

Project 5.5   Binary Equivalent of a Decimal Number

Program P5_5 can be used to the generate the binary equivalent in sign-magnitude form of
a decimal fraction.

```
% Program P5_5
% Determines the binary equivalent of a
% decimal number in sign-magnitude form
d = input('Type in the decimal fraction = ');
b = input('Type in the desired wordlength = ');
d1 = abs(d);
beq = [zeros(1,b)];
for k = 1:b
    int = fix((2*d1);
    beq(k) = int;
    d1 = 2*d1 - int;
end
if sign(d) == -1;
    bin = [1 beq];
else
    bin = [0 beq];
end
disp('The binary equivalent is');
disp(bin)
```

Questions:

Q5.18   What is the function of the operator == in Programs P5_5?

Q5.19   Using Program P5_5 develop the binary equivalents in sign-magnitude form of the
following decimal fractions: (a) 0.80165, (b) − 0.80165, (c) 0.64333, and (d) − 0.9125 for
the following values of the wordlengths: 6 and 8. Verify the results by hand calculation.

Project 5.6   Decimal Equivalent of a Binary Number

Program P5_6 performs the reverse process and generates the decimal equivalent of a binary fraction in sign-magnitude form.

```
% Program P5_6
% Determines the decimal equivalent of a
% binary number in sign-magnitude form
bin = input('Type in the binary fraction = ');
b = length(bin) - 1; d = 0;
for k = 1:b
    d = d + bin(k+1)*2^(-k);
end
if sign(bin(1)) == 0;
```

```
    dec = d;
else
    dec = - d;
end
disp('The decimal equivalent is');
disp(dec);
```

Question:

Q5.20 Using Program P5_6 determine the decimal equivalents of the binary fractions developed in Question Q5.19. How close are your answers to the original decimal fractions?

Project 5.7   Binary Number Representation Schemes

Program P5_7 can be used to determine the ones'-complement of a binary number in sign-magnitude form , whereas Program P5_8 can be used to determine the two's-complement representation of a negative binary fraction in ones'-complement form.

```
% Program P5_7
% Determines the ones'-complement equivalent of a
% binary number in sign-magnitude form
bin = input('Type in the binary number = ');
if sign(bin(1)) == 0;
    onescomp = bin;
else
    bin(1) = 0;onescomp = ~bin;
end
disp('Ones-complement equivalent is');
disp(onescomp);

% Program P5_8
% Determines the two's-complement equivalent of a
% negative binary fraction in ones'-complement form
b = input('Type in the binary fraction = ');
F = length(b);
twoscomp = ones(1,F);
c = 1;
for k = F:-1:2
    if b(k) & c == 1;
       twoscomp(k) = 0; c = 1;
       else
       twoscomp(k) = b(k) | c; c = 0;
    end
end
disp('Twos-complement equivalent is = ');
disp(twoscomp)
```

Questions:

Q5.21   What is the purpose of the operator ~ in Program P5_7?

Q5.22   Using Program P5_7 determine and verify the ones'-complement representations of the binary numbers developed in Question Q5.19.

Q5.23   What are the purposes of the operators | and & in Program P5_8?

Q5.24   Using Program P5_8 determine and verify the two's-complement representations of the binary numbers developed in Question Q5.19.

Project 5.8   D/A Converter Droop Compensation

In this project you will investigate the droop compensation by means of a digital filter inserted before the D/A converter . Two very simple low-order droop compensation digital filters are characterized by the transfer functions [Jac96]:

$$H_{FIR}(z) = \frac{1}{16}(-1 + 18\,z^{-1} - z^{-2}), \tag{5.29}$$

$$H_{IIR}(z) = \frac{9}{8 + z^{-1}}. \tag{5.30}$$

Question:

Q5.25   Write a MATLAB program to determine and plot the magnitude responses of the uncompensated and the droop-compensated D/A converters in the same figure. Use both the FIR and the IIR droop compensation filters of Eqs. (5.29) and (5.30). Run this program and comment on your results.

5.8   Background Reading

[1] A. Antoniou. *Digital Filters: Analysis, Design, and Applications*. McGraw-Hill, New York NY, second edition,1993. Chs. 5, 6.

[2] E. Cunningham. *Digital Filtering: An Introduction*. Houghton-Mifflin, Boston MA, 1992. Secs. 2.4, 3.2, 3.3.

[3] D.J. DeFatta, J.G. Lucas, and W.S. Hodgkiss. *Digital Signal Processing: A System Design Approach*. Wiley, New York NY, 1988. Secs. 2.5, 4.2.

[4] L.B. Jackson. *Digital Filters and Signal Processing*. Kluwer, Boston MA, third edition, 1996. Secs. 6.3, 8.1.

[5] R. Kuc. *Introduction to Digital Signal Processing*, McGraw-Hill, New York, NY, 1988, Secs. 3.10, 3.11.

[6] L.C. Ludeman, *Fundamentals of Digital Signal Processing*. Harper & Row, New York NY, 1986. Secs. 1.5, 1.6 and Ch. 3.

[7] S.K. Mitra. *Digital Signal Processing: A Computer-Based Approach*. McGraw-Hill, New York NY, 1998. Ch. 5 and Sec. 8.4.

[8] A.V. Oppenheim and R.W. Schafer. *Discrete-Time Signal Processing*. Prentice-Hall, Englewood Cliffs NJ, 1989. Secs. 3.1–3.3, Appendix B.

[9] S.J. Orfanidis. *Introduction to Signal Processing*. Prentice-Hall, Englewood Cliffs NJ, 1996. Ch. 1.

[10] B. Porat. *A Course in Digital Signal Procesing*. Wiley, New York NY, 1996. Secs. 3.1–3.5, 38. 10.1–10.4.

[11] J.G. Proakis and D.G. Manolakis. *Digital Signal Processing: Principles, Algorithms, and Applications*. Prentice-Hall, Englewood Cliffs NJ, third edition, 1996. Secs. 1.4.1, 1.4.2, 8.3.5, 9.1.

[12] R.A. Roberts and C.T. Mullis. *Digital Signal Processing*. Addison-Wesley, Reading MA, 1987. Secs. 4.6, 4.7.

Digital Filter Structures  6

6.1 Introduction

A structural representation using interconnected basic building blocks is the first step in the hardware or software implementation of an LTI digital filter. The structural representation provides the relations between some pertinent internal variables with the input and the output that in turn provide the keys to the implementation. This exercise considers the development of structural representations of causal IIR and FIR transfer functions in the form of block diagrams.

6.2 Background Review

R6.1 The computational algorithm of an LTI digital filter can be conveniently represented in a block-diagram form using the basic building blocks representing the unit delay, the multiplier, the adder, and the pick-off node as depicted in Figure 6.1.

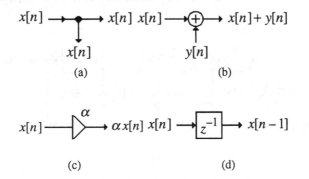

Figure 6.1 Basic building blocks: (a) pick-off node, (b) adder, (c) multiplier, and (d) unit delay.

R6.2 Two digital filter structures are called *equivalent* if they have the same transfer function. A fairly simple way to generate an equivalent structure from a given realization is via the *transpose* operation which is as follows: (i) Reverse all paths, (ii) replace pick-off nodes by adders and vice-versa, and (iii) interchange the input and the output nodes.

R6.3 Structures in which the multiplier coefficients are precisely the coefficients of the transfer function are called *direct form* structures.

R6.4   A causal FIR filter of length M is characterized by a transfer function $H(z)$:

$$H(z) = \sum_{k=0}^{M-1} h[k]\, z^{-k}, \tag{6.1}$$

which is a polynomial in z^{-1} of degree $M-1$. In the time-domain the input-output relation of the above FIR filter is given by

$$y[n] = \sum_{k=0}^{M-1} h[k]\, x[n-k], \tag{6.2}$$

where $y[n]$ and $x[n]$ are the output and input sequences, respectively.

R6.5   A direct form realization of an FIR filter can be readily developed from Eq. (6.2) as indicated in Figure 6.2(a) for $M = 5$. Its transpose, as sketched in Figure 6.2(b), is the second direct form structure. An FIR filter of length M is characterized by M coefficients and, in general, requires M multipliers and $(M - 1)$ two-input adders for implementation.

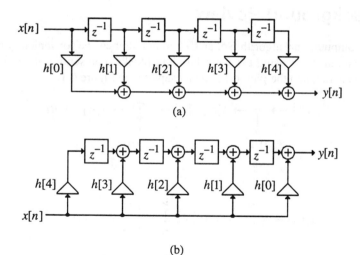

Figure 6.2   Direct form FIR structures.

R6.6   A higher order FIR transfer function can also be realized as a cascade of FIR sections with each section characterized by either a first-order or a second order transfer function. To this end, the FIR transfer function $H(z)$ of Eq. (6.1) is expressed in a factored form as

$$H(z) = h[0] \prod_{k} (1 + \beta_{1k}z^{-1} + \beta_{2k}z^{-2}), \tag{6.3}$$

where for a first-order factor $\beta_{2k} = 0$. A realization of Eq. (6.3) is shown in Figure 6.3 for a cascade of three second-order sections. Each second-order stage in Figure 6.3, of course, can be realized also in the transposed direct form. The cascade form realization also requires, in general, $(M - 1)$ two-input adders and M multipliers for an FIR transfer function of length M.

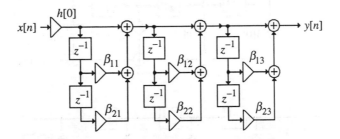

Figure 6.3 Cascade form FIR structure for a length-7 FIR filter.

R6.7 A length-M linear-phase FIR filter is characterized by either a symmetric impulse response $h[n] = h[M - 1 - n]$ or an antisymmetric impulse response $h[n] = -h[M - 1 - n]$. The symmetry (or the antisymmetry) property of a linear-phase FIR filter can be exploited to reduce the total number of multipliers into half of those needed in the direct form implementations of the transfer function. For example, Figure 6.4(a) shows the realization of a length-7 Type 1 FIR transfer function with a symmetric impulse response and Figure 6.4(b) shows the realization of a length-8 Type 2 FIR transfer function with a symmetric impulse response.

R6.8 A causal IIR filter of order N is characterized by a transfer function $H(z)$:

$$H(z) = \frac{\sum_{k=0}^{N} p_k \, z^{-k}}{1 + \sum_{k=1}^{N} d_k \, z^{-k}}, \qquad (6.4)$$

which is a ratio of polynomials in z^{-1} of degree N. In the time-domain the input-output relation of the above IIR filter is given by

$$y[n] = \sum_{k=0}^{N} p_k \, x[n - k] - \sum_{k=1}^{N} d_k \, y[n - k], \qquad (6.5)$$

where $y[n]$ and $x[n]$ are the output and input sequences, respectively.

R6.9 By defining an intermediate signal variable $w[n]$:

$$w[n] = \sum_{k=0}^{N} p_k \, x[n - k], \qquad (6.6)$$

the difference equation of Eq. (6.5) can be alternately written as

$$y[n] = w[n] - \sum_{k=1}^{N} d_k \, y[n - k]. \qquad (6.7)$$

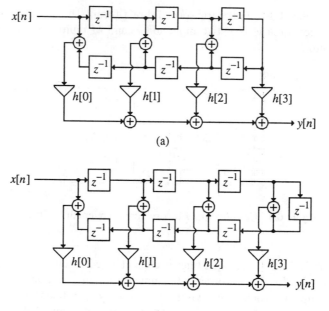

Figure 6.4   Linear-phase FIR structures: (a) Type 1 and (b) Type 2.

A realization of the IIR filter based on Eqs. (6.6) and (6.7) is called a *Direct Form I* structure and is shown in Figure 6.5(a) for $N = 3$. Its transposed form is shown in Figure 6.5(b). The total number of delays required in Direct Form I realization is $2N$ which can be reduced to N by simple block-diagram manipulations resulting in *Direct Form II* structures indicated in Figure 6.6 for $N = 3$.

An Nth order IIR digital filter transfer function is characterized by $2N + 1$ unique coefficients and, in general, requires $2N + 1$ multipliers and $2N$ two-input adders for implementation.

R6.10   By expressing the numerator and the denominator polynomials of the transfer function $H(z)$ as a product of polynomials of lower degree, a digital filter is often realized as a cascade of low-order filter sections. Usually, the polynomials are factored into a product of first-order and second-order polynomials. In this case, $H(z)$ is expressed as

$$H(z) = p_0 \prod_k \left(\frac{1 + \beta_{1k}z^{-1} + \beta_{2k}z^{-2}}{1 + \alpha_{1k}z^{-1} + \alpha_{2k}z^{-2}} \right). \tag{6.8}$$

In the above, for a first-order factor $\alpha_{2k} = \beta_{2k} = 0$. A possible realization of a third-order transfer function

$$H(z) = p_0 \left(\frac{1 + \beta_{11}z^{-1}}{1 + \alpha_{11}z^{-1}} \right) \left(\frac{1 + \beta_{12}z^{-1} + \beta_{22}z^{-2}}{1 + \alpha_{12}z^{-1} + \alpha_{22}z^{-2}} \right), \tag{6.9}$$

is shown in Figure 6.7.

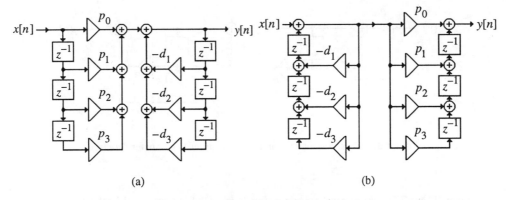

Figure 6.5   (a) Direct Form I structure and (b) transposed Direct Form I structure.

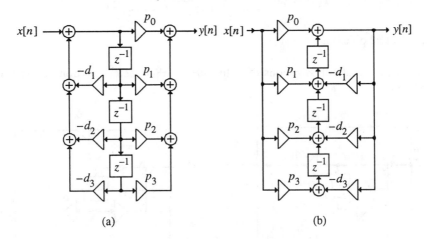

Figure 6.6   (a) Direct Form II structure and (b) transposed Direct Form II structure.

R6.11   An IIR transfer function can be realized in a parallel form by making use of the partial-fraction expansion of the transfer function. A partial-fraction expansion of the transfer function in z^{-1} leads to the *Parallel Form I*. Thus, assuming simple poles, $H(z)$ is expressed in the form

$$H(z) = \gamma_0 + \sum_k \left(\frac{\gamma_{0k} + \gamma_{1k}z^{-1}}{1 + \alpha_{1k}z^{-1} + \alpha_{2k}z^{-2}} \right). \qquad (6.10)$$

In the above, for a real pole $\alpha_{2k} = \gamma_{1k} = 0$.

A direct partial-fraction expansion of the transfer function $H(z)$ expressed as a ratio of polynomials in z, leads to the second basic form of the parallel structure, called the *Paral-*

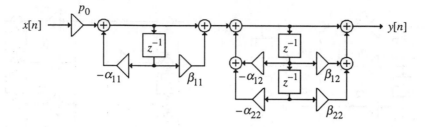

Figure 6.7 Cascade realization of a third-order IIR transfer function.

lel Form II [Mit77a]. Assuming simple poles, here $H(z)$ is expressed in the form

$$H(z) = \delta_0 + \sum_k \left(\frac{\delta_{1k}z^{-1} + \delta_{2k}z^{-2}}{1 + \alpha_{1k}z^{-1} + \alpha_{2k}z^{-2}} \right). \tag{6.11}$$

Here, for a real pole $\alpha_{2k}k = \delta_{2k} = 0$.

The two basic parallel realizations of a third-order IIR transfer function are sketched in Figure 6.8.

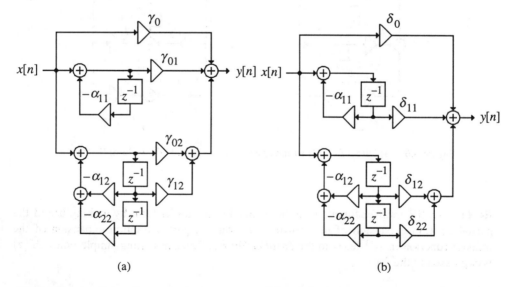

Figure 6.8 Parallel realizations of a third-order IIR transfer function: (a) Parallel Form I and (b) Parallel Form II.

R6.12 An M-th order real coefficient allpass transfer function

$$A_M(z) = \frac{d_M + d_{M-1}z^{-1} + \ldots + d_1z^{-(M-1)} + z^{-M}}{1 + d_1z^{-1} + \ldots + d_{M-1}z^{-(M-1)} + d_Mz^{-M}}, \tag{6.12}$$

is characterized by M unique coefficients and can be realized using only M multipliers. In one method, $A_M(z)$ is realized in the form of a cascade of second-order and first-order allpass sections. In the second method, $A_M(z)$ is realized as a first-order lattice two-pair constrained by an allpass transfer function $A_{M-1}(z)$ of order $M-1$. By repeating the process, a realization of $A_M(z)$ is obtained in the form of a cascaded lattice structure.

R6.13   Two one-multiplier realizations of a first-order allpass transfer function

$$A_1(z) = \frac{d_1 + z^{-1}}{1 + d_1 z^{-1}},$$ (6.13)

called *Type 1 allpass structures*, are shown in Figure 6.9 [Mit74a]. Transpose of these structures yields two other Type 1 allpass structures.

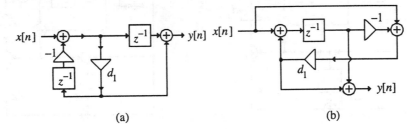

(a)   (b)

Figure 6.9   (a) Type 1A allpass structure and (b) Type 1B allpass structure.

R6.14   Two-multiplier realizations of the second-order allpass transfer function of the form

$$A_2(z) = \frac{d_1 d_2 + d_1 z^{-1} + z^{-2}}{1 + d_1 z^{-1} + d_1 d_2 z^{-2}},$$ (6.14)

called *Type 2 allpass structures*, are shown in Figure 6.10 [Mit74a]. Additional Type 2 allpass structures can be derived by transposing these structures.

R6.15   Two-multiplier realizations of the second-order allpass transfer function of the form

$$A_2(z) = \frac{d_2 + d_1 z^{-1} + z^{-2}}{1 + d_1 z^{-1} + d_2 z^{-2}},$$ (6.15)

called *Type 3 allpass structures*, are shown in Figure 6.11 [Mit74a]. Additional Type 3 allpass structures can be derived by transposing these structures.

R6.16   The transfer functions of the allpass structures of Figures 6.9 - 6.11 remain allpass for any values of the multiplier coefficients, and are called *structurally lossless bounded-real* (LBR) as long as they are stable.

R6.17   The cascaded lattice realization of an Mth order allpass transfer function is based on the development of a series of $(m-1)$th order allpass transfer functions $A_{m-1}(z)$ from an mth order allpass transfer function $A_m(z)$, $m = M, M-1, \ldots, 1$ [Vai87]:

$$A_m(z) = \frac{d_m + d_{m-1} z^{-1} + d_{m-2} z^{-2} + \ldots + d_1 z^{-(m-1)} + z^{-m}}{1 + d_1 z^{-1} + d_2 z^{-2} + \ldots + d_{m-1} z^{-(m-1)} + d_m z^{-m}},$$ (6.16)

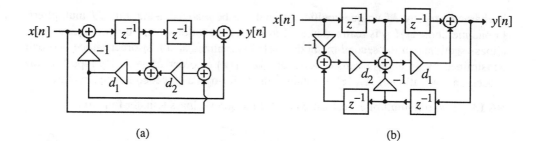

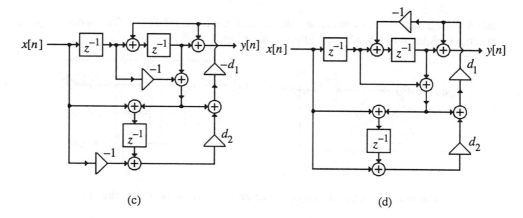

Figure 6.10 (a) Type 2A allpass structure, (b) Type 2D allpass structure, (c) Type 2B allpass structure, and (d) Type 2C allpass structure.

using the recursion

$$A_{m-1}(z) = z\left[\frac{A_m(z) - k_m}{1 - k_m A_m(z)}\right], \qquad m = M, M-1, \ldots, 1, \qquad (6.17)$$

where $k_m = A_m(\infty) = d_m$. It can be shown that $A_M(z)$ is stable if and only if

$$k_m^2 < 1, \qquad \text{for} \qquad m = M, M-1, \ldots, 1. \qquad (6.18)$$

If the allpass transfer function $A_{m-1}(z)$ is expressed in the form

$$A_{m-1}(z) = \frac{d'_{m-1} + d'_{m-2}z^{-1} + \ldots + d'_1 z^{-(m-2)} + z^{-(m-1)}}{1 + d'_1 z^{-1} + \ldots + d'_{m-2}z^{-(m-2)} + d'_{m-1}z^{-(m-1)}}, \qquad (6.19)$$

then the coefficients of $A_{m-1}(z)$ are simply related to the coefficients of $A_m(z)$ through the expression:

$$d'_i = \frac{d_i - d_m d_{m-i}}{1 - d_m^2}, \qquad i = m-1, m-2, \ldots, 2, 1. \qquad (6.20)$$

A realization of $A_m(z)$ based on the recursion of Eq. (6.17) is shown in Figure 6.12(a). The cascaded lattice realization of $A_M(z)$ based on this recursion is thus as shown in Figure 6.12(b).

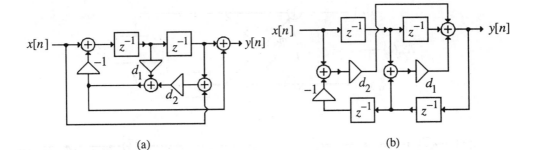

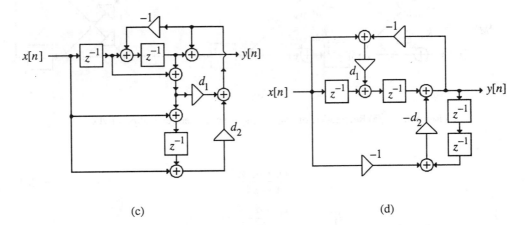

Figure 6.11   (a) Type 3A allpass structure, (b) Type 3D allpass structure, (c) Type 3C structure, and (d) Type 3H allpass structure.

R6.18   The cascaded lattice structure of Figure 6.12 forms the basis of the Gray-Markel method for the realization of an arbitrary Mth order transfer function $H(z)$ [Gra73]. In this method, $H(z) = P_M(z)/D_M(z)$ is realized in two steps. In the first step, an intermediate allpass transfer function $A_M(z) = z^{-M} D_M(z^{-1})/D_M(z)$ is realized in the form of a cascaded lattice structure. The state variables of this structure are then summed in the second step with appropriate weights to yield the desired numerator $P_M(z)$.

To illustrate the method of realizing the numerator, consider for simplicity the implementation of a third-order IIR transfer function

$$H(z) = \frac{P_3(z)}{D_3(z)} = \frac{p_0 + p_1 z^{-1} + p_2 z^{-2} + p_3 z^{-3}}{1 + d_1 z^{-1} + d_2 z^{-2} + d_3 z^{-3}}. \qquad (6.21)$$

To this end, first the allpass function $A_3(z) = Y_1(z)/X_1(z) = z^{-3} D_3(z^{-1})/D_3(z)$ is realized as shown in Figure 6.13(a) where

$$d_1' = \frac{d_1 - d_3 d_2}{1 - d_3^2},$$

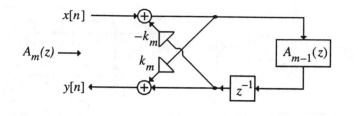

(a)

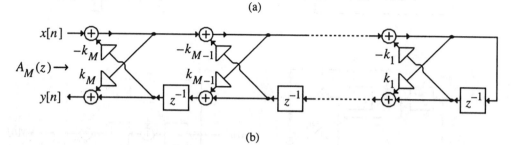

(b)

Figure 6.12 (a) Realization of $A_m(z)$ and (b) cascaded realization of $A_M(z)$.

$$d_2' = \frac{d_2 - d_3 d_1}{1 - d_3^2}, \tag{6.22}$$

$$d_1'' = \frac{d_1' - d_2' d_1'}{1 - (d_2')^2} = \frac{d_1'}{1 + d_2'}.$$

Next the signal variables Y_1, S_1, S_2, and S_3, are summed with weights $\{\alpha_i\}$ as shown in Figure 6.13(b) to arrive at the desired numerator $P_3(z)$. The weights $\{\alpha_i\}$ are given by

$$
\begin{aligned}
\alpha_1 &= p_3, \\
\alpha_2 &= p_2 - \alpha_1 d_1, \\
\alpha_3 &= p_1 - \alpha_1 d_2 - \alpha_2 d_1', \\
\alpha_3 &= p_0 - \alpha_1 d_3 - \alpha_2 d_2' - \alpha_3 d_1''.
\end{aligned} \tag{6.23}
$$

R6.19 Let $G(z)$ be an Nth order causal bounded-real IIR transfer function having a symmetric numerator and let $H(z)$ be the Nth order power-complementary causal bounded-real transfer function of $G(z)$ with an antisymmetric numerator. Then $G(z)$ and $H(z)$ can always be decomposed in the form

$$G(z) = \frac{1}{2}\{A_0(z) + A_1(z)\},$$

$$H(z) = \frac{1}{2}\{A_0(z) - A_1(z)\}. \tag{6.24}$$

where $A_0(z)$ and $A_1(z)$ are causal stable allpass transfer functions with the sum of their degrees being N [Vai86]. The realization of $G(z)$ and $H(z)$ based on the above decomposition is thus as shown in Figure 6.14.

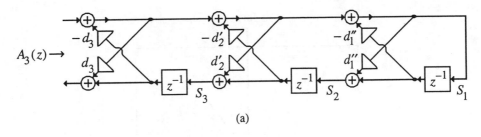

(a)

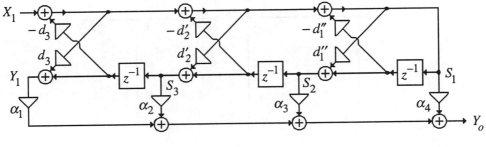

(b)

Figure 6.13 (a) Cascaded lattice realization of a third-order allpass transfer function and (b) Gray-Markel realization of $H(z)$ of Eq. (6.21).

In the case of odd-order digital Butterworth, Chebyshev, and elliptic lowpass or highpass digital transfer functions (discussed in Chapter 7), there is a simple approach to identify the poles of the allpass transfer functions $A_0(z)$ and $A_1(z)$ from the poles $\lambda_k, 0 \le k \le N-1$, of the parent lowpass transfer function $G(z)$ or $H(z)$. Let θ_k denote the angle of the pole λ_k. If we assume that the poles are numbered such that $\theta_k < \theta_{k+1}$, then the poles of $A_0(z)$ are given by θ_{2k} and the poles of $A_1(z)$ are given by θ_{2k+1} [Gas85]. Figure 6.15 illustrates this pole interlacing property of the two allpass transfer functions. Zeros of the allpass transfer functions are situated at the mirror-image locations with respect to their pole locations.

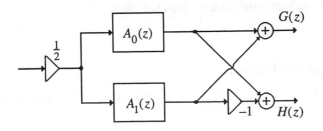

Figure 6.14 Parallel allpass realization of an IIR transfer function.

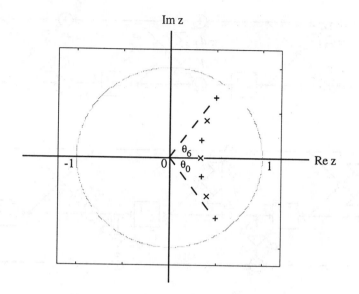

Figure 6.15 Illustration of pole interlacing property. The poles marked + belong to $A_0(z)$ and the poles marked × belong to $A_1(z)$

6.3   MATLAB Commands Used

The MATLAB comands you will encounter in this exercise are as follows:

General Purpose Commands

    `disp`   `length`

Operators and Special Characters

    `:`   `.`   `+`   `-`   `*`   `/`

    `;`   `%`

Elementary Matrices and Matrix Manipulation

    `ones`   `pi`   `:`

Signal Processing Toolbox

    `latc2tf`   `poly2rc`   `residue`   `residuez`
    `tf2latc`   `zp2sos`

For additional information on these commands, see the *MATLAB Reference Guide* [Mat94] and the *Signal Processing Toolbox User's Guide* [Mat96] or type `help commandname` in the Command window. A brief explanation of the MATLAB functions used here can be found in Appendix B.

6.4   Realization of FIR Transfer Functions

Project 6.1   Cascade Realization

The factored form of a causal FIR transfer function $H(z)$ of order $M - 1$, as given in Eq. (6.3) can be determined from its polynomial form representation given by Eq. (6.1) which can then be utilized to realize $H(z)$ in a cascade form. To this end, a modified form of Program P6_1 that uses the function zp2sos can be employed.

```
% Program P6_1
% Conversion of a rational transfer function
% to its factored form
num = input('Numerator coefficient vector = ');
den = input('Denominator coefficient vector = ');
[z,p,k] = tf2zp(num,den);
sos = zp2sos(z,p,k)
```

Questions:

Q6.1   Using Program P6_1 develop a cascade realization of the following FIR transfer function:

$$H_1(z) = 2 + 10\,z^{-1} + 23\,z^{-2} + 34\,z^{-3} + 31\,z^{-4} + 16\,z^{-5} + 4\,z^{-6}. \qquad (6.25)$$

Sketch the block diagram of the cascade realization. Is $H_1(z)$ a linear-phase transfer function?

Q6.2   Using Program P6_1 develop a cascade realization of the following FIR transfer function:

$$H_2(z) = 6 + 31\,z^{-1} + 74\,z^{-2} + 102\,z^{-3} + 74\,z^{-4} + 31\,z^{-5} + 6\,z^{-6}. \qquad (6.26)$$

Sketch the block diagram of the cascade realization. Is $H_2(z)$ a linear-phase transfer function? Develop a cascade realization of $H_2(z)$ with only 4 multipliers. Show the block-diagram of the new cascade structure.

6.5   Realization of IIR Transfer Functions

Project 6.2   Cascade and Parallel Realizations

The factored form of a causal IIR transfer function $H(z)$ of order N as given in Eq. (6.8) can be determined from its rational form representation given by Eq. (6.4), which then can be used to realize $H(z)$ in a cascade form. To this end, Program P6_1 can be employed.

Questions:

Q6.3   Using Program P6_1 develop a cascade realization of the following causal IIR transfer function:

$$H_1(z) = \frac{3 + 8\,z^{-1} + 12\,z^{-2} + 7\,z^{-3} + 2\,z^{-4} - 2\,z^{-5}}{16 + 24\,z^{-1} + 24\,z^{-2} + 14\,z^{-3} + 5\,z^{-4} + z^{-5}}. \tag{6.27}$$

Sketch the block diagram of the cascade realization.

Q6.4   Using Program P6_1 develop a cascade realization of the following causal IIR transfer function:

$$H_2(z) = \frac{2 + 10\,z^{-1} + 23\,z^{-2} + 34\,z^{-3} + 31\,z^{-4} + 16\,z^{-5} + 4\,z^{-6}}{36 + 78\,z^{-1} + 87\,z^{-2} + 59\,z^{-3} + 26\,z^{-4} + 7\,z^{-5} + z^{-6}}. \tag{6.28}$$

Sketch the block diagram of the cascade realization.

There are two parallel form realizations of a causal IIR transfer function. Parallel Form I is based on its partial-fraction expansion in z^{-1} as in Eq. (6.10), which can be obtained using MATLAB function `residuez`. Parallel Form II is based on the partial-fraction expansion in z as in Eq. (6.11), which is obtained using the function `residue`. Program P6_2 develops both types of parallel realizations.

```
% Program P6_2
% Parallel Form Realizations of an IIR Transfer Function
num = input('Numerator coefficient vector = ');
den = input('Denominator coefficient vector = ');
[r1,p1,k1] = residuez(num,den);
[r2,p2,k2] = residue(num,den);
disp('Parallel Form I')
disp('Residues are');disp(r1);
disp('Poles are at');disp(p1);
disp('Constant value');disp(k1);
disp('Parallel Form II')
disp('Residues are');disp(r2);
disp('Poles are at');disp(p2);
disp('Constant value');disp(k2);
```

Questions:

Q6.5   Using Program P6_2 develop the two different parallel-form realizations of the causal IIR transfer function of Eq. (6.27). Sketch the block diagrams of both realizations.

Q6.6   Using Program P6_2 develop the two different parallel-form realizations of the causal IIR transfer function of Eq. (6.28). Sketch the block diagrams of both realizations.

Project 6.3   Realization of an Allpass Transfer Function

The cascaded lattice realization of an Mth order causal IIR allpass transfer function $A_M(z)$ is based on the recursive algorithm outlined in R6.17. The lattice parameters $\{k_i\}$ can be determined in MATLAB using the function poly2rc. To this end, Program P4_4 can also be employed.

Questions:

Q6.7 Using Program P4_4 develop the cascaded lattice realization of the following all-pass transfer function:

$$A_5(z) = \frac{1 + 5\,z^{-1} + 14\,z^{-2} + 24\,z^{-3} + 24\,z^{-4} + 16\,z^{-5}}{16 + 24\,z^{-1} + 24\,z^{-2} + 14\,z^{-3} + 5\,z^{-4} + z^{-5}}. \tag{6.29}$$

Is $A_5(z)$ a stable transfer function?

Q6.8 Using Program P4_4 develop the cascaded lattice realization of the following all-pass transfer function:

$$A_6(z) = \frac{1 + 7\,z^{-1} + 26\,z^{-2} + 59\,z^{-3} + 87\,z^{-4} + 78\,z^{-5} + 36\,z^{-6}}{36 + 78\,z^{-1} + 87\,z^{-2} + 59\,z^{-3} + 26\,z^{-4} + 7\,z^{-5} + z^{-6}}. \tag{6.30}$$

Is $A_6(z)$ a stable transfer function?

A higher-order allpass transfer function can also be realized as a cascade of second-order and first-order allpass sections described in R6.13 – R6.15 using a modified form of Program 6_1. It should be noted that the pairing of numerator and denominator factors obtained via zp2sos does not result in allpass sections. However, it is easy to determine the denominator factors by forming the mirror-image factors of the numerator decomposition.

Questions:

Q6.9 Develop a canonic cascade realization of the allpass transfer function of Eq. (6.29) using Types 1 and 2 allpass sections. Show the block diagram of the realization. What is the total number of multipliers in the final structure?

Q6.10 Develop a canonic cascade realization of the allpass transfer function of Eq. (6.30) using three allpass sections. Show the block diagram of the realization. What is the total number of multipliers in the final structure?

Project 6.4   Gray-Markel Realization of an IIR Transfer function

The Gray-Markel cascaded lattice realization of an Nth order causal IIR transfer function $H(z)$ is based on the cascaded lattice realization of an intermediate allpass transfer function $A_N(z)$ with the same denominator as that of $H(z)$ followed by a weighted combination of the internal state-variables and the allpass output variable as outlined in R6.17 and R6.18. Program P6_3 implements this algorithm.

```
% Program P6_3
% Gray-Markel Cascaded Lattice Structure
% k is the lattice parameter vector
% alpha is the vector of feedforward multipliers
format long
% Read in the transfer function coefficients
num = input('Numerator coefficient vector = ');
den = input('Denominator coefficient vector = ');
N = length(den)-1; % Order of denominator polynomial
k = ones(1,N);
a1 = den/den(1);
alpha = num(N+1:-1:1)/den(1);
for ii = N:-1:1,
    alpha(N+2-ii:N+1) = alpha(N+2-ii:N+1)-alpha(N-ii+1)*a1(2:ii+1);
    k(ii) = a1(ii+1);
    a1(1:ii+1) = (a1(1:ii+1)-k(ii)*a1(ii+1:-1:1))/(1-k(ii)*k(ii));
end
disp('Lattice parameters are');disp(k)
disp('Feedforward multipliers are');disp(alpha)
```

Questions:

Q6.11 Using Program P6_3 develop the Gray-Markel realization of the causal IIR transfer function of Eq. (6.27). Sketch the block diagrams of both realizations. Is the transfer function stable?

Q6.12 Using Program P6_3 develop the Gray-Markel realization of the causal IIR transfer function of Eq. (6.28). Sketch the block diagrams of both realizations. Is the transfer function stable?

The function tf2latc in the Signal Processing Toolbox can also be used to develop the Gray-Markel realization of a causal IIR transfer function. The basic form of this function is [k, alpha] = tf2latc(num, den) where num and den are the vectors of the coefficients of the numerator polynomial and the denominator polynomial of the transfer function in ascending powers of z^{-1}, respectively. All coefficients must be normalized by the leading coefficient of den. The output data are the vector k of the lattice parameters and the vector alpha of the feedforward multiplier coefficients. It should be noted that the ordering of the feedforward coefficients is exactly opposite to that generated by Program P6_3. It is also possible to determine the transfer function from the lattice parameter vector k and the feedforward coefficients vector alpha using the function latc2tf. In this case the statement to use is [num, den] = latc2tf(k, alpha).[1]

---

[1]The function latc2tf in the Signal Processing Toolbox should be modified to make it work properly. The suuggested corrections can be downloaded via anonymous ftp from **ftp://ftp.mathworks.com/pub/tech-support/signal/latc2tf.m**.

Questions:

Q6.13 Write a MATLAB program using the function `tf2latc` to develop the Gray-Markel realization of a causal IIR transfer function. Using this program realize the transfer function of Eq. (6.27). Does your result check with that obtained in Question 6.11? Using the function `latc2tf` determine the transfer function from the vectors k and `alpha`. Is the transfer function obtained the same as in Eq. (6.27)?

Q6.14 Using the program developed in Question Q6.13 realize the transfer function of Eq. (6.28). Does your result check with that obtained in Question Q6.12? Using the function `latc2tf` determine the transfer function from the vectors k and `alpha`. Is the transfer function obtained the same as in Eq. (6.28)?

Project 6.5   Parallel Allpass Realization of an IIR Transfer Function

Questions:

Q6.15 Develop the sum-of-allpass decomposition of a third order causal bounded-real lowpass Type 1 Chebyshev transfer function $G(z)$ given by

$$G(z) = \frac{0.0736 + 0.2208\,z^{-1} + 0.2208\,z^{-2} + 0.0736\,z^{-3} + 87\,z^{-4}}{1 - 0.9761\,z^{-1} + 0.8568\,z^{-2} - 0.2919\,z^{-3}}. \tag{6.31}$$

What is the expression for its power-complementary transfer function $H(z)$? What are the orders of the two allpass transfer functions? Develop the parallel allpass realization of $G(z)$ and $H(z)$ with at most three multipliers by realizing the two allpass transfer functions as a cascade of first-order and/or second-order sections.

Q6.16 Develop the sum-of-allpass decomposition of a fifth order causal bounded-real lowpass elliptic transfer function $G(z)$ where

$$G(z) = \frac{0.0417 + 0.07675\,z^{-1} + 0.1203\,z^{-2} + 0.1203\,z^{-3} + 0.0767\,z^{-4} + 0.0417\,z^{-5}}{1 - 1.8499\,z^{-1} + 2.5153\,z^{-2} - 1.9106\,z^{-3} + 0.9565\,z^{-4} - 0.234\,z^{-5}}. \tag{6.32}$$

What is the expression for its power-complementary transfer function $H(z)$? What are the orders of the two allpass transfer functions? Develop the parallel allpass realization of $G(z)$ and $H(z)$ with at most five multipliers by realizing the two allpass transfer functions as a cascade of first-order and/or second-order sections.

6.6   Background Reading

[1] A. Antoniou., *Digital Filters: Analysis, Design, and Applications*. McGraw-Hill, New York NY, second edition,1993. Ch. 4.

[2] E. Cunningham. *Digital Filtering: An Introduction*. Houghton-Mifflin, Boston MA, 1992. Sec. 3.12.

[3] D.J. DeFatta, J.G. Lucas, and W.S. Hodgkiss. *Digital Signal Processing: A System Design Approach*. Wiley, New York NY, 1988. Sec. 3.5.

[4] L.B. Jackson. *Digital Filters and Signal Processing*. Kluwer, Boston MA, third edition, 1996. Ch. 5.

[5] R. Kuc. *Introduction to Digital Signal Processing*. McGraw-Hill, New York NY, 1988. Ch. 6.

[6] L.C. Ludeman. *Fundamentals of Digital Signal Processing*. Harper & Row, New York NY, 1986. Ch. 5.

[7] S.K. Mitra. *Digital Signal Processing: A Computer-Based Approach*. McGraw-Hill, New York NY, 1998. Ch. 6.

[8] A.V. Oppenheim and R.W. Schafer. *Discrete-Time Signal Processing*. Prentice-Hall, Englewood Cliffs NJ, 1989. Secs. 6.1 – 6.6.

[9] S.J. Orfanidis. *Introduction to Signal Processing*. Prentice-Hall, Englewood Cliffs NJ, 1996. Secs. 4.2.2, 7.1 – 7.3.

[10] B. Porat. *A Course in Digital Signal Procesing*. Wiley, New York NY, 1996. Sec. 11.1.

[11] J.G. Proakis and D.G. Manolakis. *Digital Signal Processing: Principles, Algorithms, and Applications*. Prentice-Hall, Englewood Cliffs NJ, third edition, 1996. Secs. 7.1 – 7.3.

[12] R.A. Roberts and C.T. Mullis. *Digital Signal Processing*. Addison-Wesley, Reading MA, 1987. Sec. 3.7.

Digital Filter Design

7.1 Introduction

The process of deriving the transfer function $G(z)$ whose frequency response $G(e^{j\omega})$ approximates the given frequency response specifications is called *digital filter design*. After $G(z)$ has been obtained, it is then realized in the form of a suitable filter structure. In the previous laboratory exercise, the realizations of FIR and IIR transfer functions have been considered. In this laboratory exercise you will learn how to design an IIR or FIR digital filter to meet a specified magnitude or gain response.

7.2 Background Review

R7.1 The filter specifications are usually specified in terms of its magnitude response. For example, the magnitude $|G(e^{j\omega})|$ of a lowpass filter $G(z)$ is usually specified as indicated in Figure 7.1. In the *passband* defined by $0 \le \omega \le \omega_p$, we require

$$1 - \delta_p \le |G(e^{j\omega})| \le 1 + \delta_p, \quad \text{for} \quad |\omega| \le \omega_p, \tag{7.1}$$

or in other words, the magnitude approximates unity within an error of $\pm\delta_p$. In the *stopband*, defined by $\omega_s \le |\omega| \le \pi$, we require

$$|G(e^{j\omega})| \le \delta_s, \quad \text{for} \quad \omega_s \le |\omega| \le \pi, \tag{7.2}$$

implying that the magnitude approximate zero within an error of δ_s. The frequencies ω_p and ω_s are, respectively, called the *passband edge frequency* and the *stopband edge frequency*.

The limits of the tolerances in the passband and stopband, δ_p and δ_s, are called *ripples*.

R7.2 In most applications, the digital filter specifications are given as indicated in Figure 7.2. Here, in the *passband* defined by $0 \le \omega \le \omega_p$, the maximum and the minimum values of the magnitude are, respectively, unity and $1/\sqrt{1 + \epsilon^2}$. The *peak passband ripple* is

$$R_p = 20 \log_{10} \sqrt{1 + \epsilon^2} \ \text{dB}. \tag{7.3}$$

The maximum stopband ripple in the *stopband*, defined by $\Omega_s \le \omega \le \pi$, is denoted by $1/A$ and the *minimum stopband attenuation* is given by

$$R_s = 20 \log_{10} A \ \text{dB}. \tag{7.4}$$

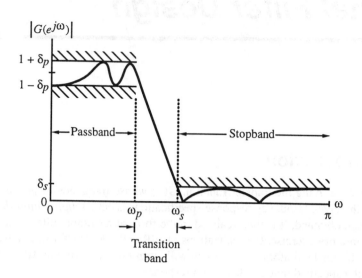

Figure 7.1   Typical magnitude response specifications for a digital lowpass filter.

R7.3   If the passband edge frequency F_p and the stopband edge frequency F_s are specified in Hz along with the sampling rate F_T of the digital filter, then the normalized angular edge frequencies in radians are given by

$$\omega_p = \frac{\Omega_p}{F_T} = \frac{2\pi F_p}{F_T} = 2\pi F_p T, \tag{7.5}$$

$$\omega_s = \frac{\Omega_s}{F_T} = \frac{2\pi F_s}{F_T} = 2\pi F_p T. \tag{7.6}$$

R7.4   The first step in the filter design process is the estimation of the order of the transfer function. For the design of an IIR digital lowpass filter $G(z)$ based on the conversion of an analog lowpass filter $H_a(s)$, an analytical formula exists for the estimation of the filter order. For the design of FIR lowpass or highpass digital filters, there are several design formulae for estimating the minimum filter length N directly from the digital filter specifications: normalized passband edge angular frequency ω_p, normalized stopband edge angular frequency ω_s, peak passband ripple δ_p, and peak stopband ripple δ_s. A rather simple approximate formula developed by Kaiser [Kai74] is given by

$$N \cong \frac{-20 \log_{10}(\sqrt{\delta_p \delta_s}) - 13}{14.6(\Delta\omega)/2\pi}, \tag{7.7}$$

where $\Delta\omega = |\omega_p - \omega_s|$ is the width of the transition band. The above formula also can be used for designing multitransition-band FIR filters, in which case $\Delta\omega$ is the width of the smallest of all transition bands. For multiband filters with unequal transition bands, the filter designed using the above estimated formula may exhibit unacceptable magnitude responses in the transition bands that are wider, in which case, these bands should be made smaller until an acceptable magnitude response is obtained.

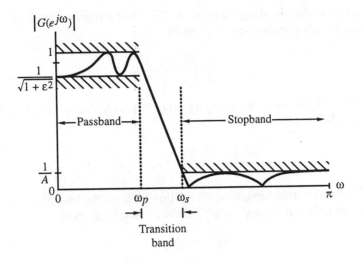

Figure 7.2 Normalized magnitude response specifications for a digital lowpass filter.

R7.5 A slightly more accurate value for the length is due to Herrmann, Rabiner, and Chan [Her73], and is given by

$$N \cong \frac{D_\infty(\delta_p, \delta_s) - F(\delta_p, \delta_s)\left[(\omega_s - \omega_p)/2\pi\right]^2}{\left[(\omega_s - \omega_p)/2\pi\right]}, \qquad (7.8)$$

where

$$\begin{aligned} D_\infty(\delta_p, \delta_s) = &\left[a_1(\log_{10}\delta_p)^2 + a_2(\log_{10}\delta_p) + a_3\right]\log_{10}\delta_s \\ &- \left[a_4(\log_{10}\delta_p)^2 + a_5(\log_{10}\delta_p) + a_6\right], \end{aligned} \qquad (7.9)$$

and

$$F(\delta_p, \delta_s) = b_1 + b_2\left[\log_{10}\delta_p - log_{10}\delta_s\right], \qquad (7.10)$$

with

$$\begin{aligned} a_1 &= 0.005309, \quad a_2 = 0.07114, \quad a_3 = -0.4761, \\ a_4 &= 0.00266, \quad a_5 = 0.5941, \quad a_6 = 0.4278, \\ &b_1 = 11.01217, \quad b_2 = 0.51244. \end{aligned} \qquad (7.11)$$

The formula given in Eq. (7.8) is valid for $\delta_p > \delta_s$. If $\delta_p < \delta_s$ then the filter length formula to be used is obtained by interchanging δ_p and δ_s in Eq. (7.8). For small values of δ_p and δ_s, both Eqs. (7.7) and (7.8) provide reasonably close and accurate results. On the other hand, when the values of δ_p and δ_s are large, Eq. (7.8) yields a more accurate value for the length.

R7.6 In many filter design problems, the order estimated using either Eq. (7.7) or Eq. (7.8) may result in filters not meeting the given specifications. In such cases, the value of N should be increased gradually until the specifications are met.

R7.7   The most widely used approach to the IIR filter design is based on the bilinear transformation from s-plane to z-plane given by

$$s = \frac{2}{T}\left(\frac{1 - z^{-1}}{1 + z^{-1}}\right). \tag{7.12}$$

Using the above transformation an analog transfer function $H_a(s)$ is converted into a digital transfer function $G(z)$ according to:

$$G(z) = H_a(s)\big|_{s = \frac{2}{T}\left(\frac{1-z^{-1}}{1+z^{-1}}\right)}. \tag{7.13}$$

R7.8   For the bilinear transformation, the relation between the imaginary axis ($s = j\Omega$) in the s-plane and the unit circle ($z = e^{j\omega}$) in the z-plane is given by

$$\Omega = \tan(\omega/2), \tag{7.14}$$

which maps the entire imaginary axis in the s-plane to the unit circle in the z-plane introducing a distortion in the frequency axis called *warping*. To develop a digital filter meeting a specified magnitude response, the analog equivalents (Ω_p and Ω_s) of the critical band-edge frequencies (ω_p and ω_s) of the digital filter are first obtained using the relation of Eq. (7.14), the analog prototype $H_a(s)$ is then designed using the prewarped critical frequencies, and $H_a(s)$ is transformed using the bilinear transformation to obtain the desired digital filter transfer function $G(z)$.

R7.9   The most straight-forward method of FIR filter design is based on windowing the ideal infinite-length impulse response $h_D[n]$ obtained by an inverse discrete-time Fourier transform of the ideal frequency response $H_D(e^{j\omega})$ by an appropriate finite-length *window function* $w[n]$. The impulse response coefficients of the final design are then given by $h[n] = h_D[n] \cdot w[n]$.

R7.10   The ideal lowpass filter of Figure 4.1(a) has a zero-phase frequency response

$$H_{LP}(e^{j\omega}) = \begin{cases} 1, & |\omega| \le \omega_c, \\ 0, & \omega_c < |\omega| \le \pi \end{cases} \tag{7.15}$$

The corresponding impulse response coefficients $h_{LP}[n]$ are given by

$$h_{LP}[n] = \frac{\sin \omega_c n}{\pi n}, \quad -\infty \le n \le \infty. \tag{7.16}$$

which is seen to be doubly infinite, not absolutely summable, and therefore unrealizable. By setting all impulse response coefficients outside the range $-M \le n \le M$ equal to zero, we arrive at a finite-length noncausal approximation of length $N = 2M + 1$ that, when shifted to the right, yields the coefficients of a causal FIR lowpass filter:

$$\hat{h}_{LP}[n] = \begin{cases} \frac{\sin \omega_c(n-M)}{\pi(n-M)}, & 0 \le n \le N - 1, \\ 0, & \text{otherwise} \end{cases} \tag{7.17}$$

It should be noted that the above expression also holds for N even in which case M is a fraction.

The impulse response coefficients $h_{HP}[n]$ of the ideal highpass filter of Figure 4.1(b) are given by

$$h_{HP}[n] = \begin{cases} 1 - \frac{\omega_c}{\pi}, & \text{for } n = 0, \\ -\frac{\sin(\omega_c n)}{\pi n}, & \text{for } |n| > 0. \end{cases} \tag{7.18}$$

The impulse response coefficients $h_{BP}[n]$ of an ideal bandpass filter of Figure 4.1(c) with cutoffs at ω_{c1} and ω_{c2} are given by

$$h_{BP}[n] = \frac{\sin \omega_{c2} n}{\pi n} - \frac{\sin \omega_{c1} n}{\pi n}, \qquad -\infty \le n \le \infty. \tag{7.19}$$

and those of an ideal bandstop filter of Figure 4.1(d) with cutoffs at ω_{c1} and ω_{c2} are given by

$$h_{BS}[n] = \begin{cases} 1 - \frac{(\omega_{c2} - \omega_{c1})}{\pi}, & \text{for } n = 0, \\ \frac{\sin(\omega_{c1} n)}{\pi n} - \frac{\sin(\omega_{c2} n)}{\pi n}, & \text{for } |n| > 0. \end{cases} \tag{7.20}$$

R7.11   All of the above design methods are for single passband or single stopband filters with two magnitude levels. However, it is quite straightforward to generalize the method to the design of multilevel FIR filters and obtain the expression for the impulse response coefficients. The zero-phase frequency response of an ideal L-band digital filter $H_{ML}(z)$ is given by

$$H_{ML}(e^{j\omega}) = A_k, \qquad \text{for } \omega_{k-1} \le \omega \le \omega_k, \qquad k = 1, 2, \ldots, L, \tag{7.21}$$

where $\omega_0 = 0$ and $\omega_L = \pi$. Figure 7.3 shows the frequency response of a typical multilevel filter. Its impulse response $h_{ML}[n]$ is given by

$$h_{ML}[n] = \sum_{\ell=1}^{L} (A_\ell - A_{\ell+1}) \cdot \frac{\sin(\omega_c n)}{\pi n}, \tag{7.22}$$

with $A_{L+1} = 0$.

R7.12   The ideal *Hilbert transformer*, also called a *90-degree phase shifter*, is characterized by a frequency response given by

$$H_{HT}(e^{j\omega}) = \begin{cases} j, & -\pi < \omega < 0, \\ -j, & 0 < \omega < \pi. \end{cases} \tag{7.23}$$

The corresponding impulse response $h_{HT}[n]$ is

$$h_{HT}[n] = \begin{cases} 0, & n = 0, \\ \frac{2 \sin^2(\pi n/2)}{\pi n}, & n \ne 0. \end{cases} \tag{7.24}$$

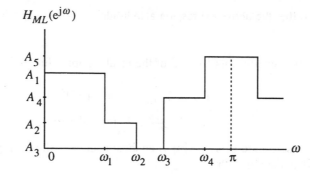

Figure 7.3   Typical zero-phase multilevel frequency response.

R7.13   The ideal *discrete-time differentiator* is characterized by a frequency response given by

$$H_{DIF}(e^{j\omega}) = j\omega, \quad |\omega| < \pi. \tag{7.25}$$

The corresponding impulse response $h_{DIF}[n]$ is

$$h_{DIF}[n] = \begin{cases} 0, & n = 0, \\ \frac{\cos \pi n}{n}, & |n| > 0. \end{cases} \tag{7.26}$$

Like the ideal lowpass filter, all of the above ideal filters (Eqs.(7.18)–(7.20), (7.22), and (7.26)) are unrealizable. They can be made realizable by truncating the impulse response sequences to finite lengths and shifting the truncated coefficients to the right appropriately.

R7.14   The causal FIR filters obtained by simply truncating the impulse response coefficients of the ideal filters given in the previous section exhibit an oscillatory behavior in their respective magnitude responses that is more commonly referred to as the *Gibbs phenomenon*. The Gibbs phenomenon can be reduced either by using a window that tapers smoothly to zero at each end or by providing a smooth transition from the passband to the stopband. Use of a tapered window causes the height of the side lobes to diminish with a corresponding increase in the main lobe width, resulting in a wider transition at the discontinuity. In all window-based lowpass filter designs, the cutoff frequency ω_c is half of the sum of the passband and stopband edge frequencies.

R7.15   Some commonly used tapered windows of length $2M + 1$ with fixed ripples are:

$$\textit{Hanning:} \quad w[n] = \frac{1}{2}\left[1 + \cos\left(\frac{2\pi n}{2M+1}\right)\right], \quad -M \leq n \leq M, \tag{7.27}$$

$$\textit{Hamming:} \quad w[n] = 0.54 + 0.46 \cos\left(\frac{2\pi n}{2M+1}\right), \quad -M \leq n \leq M, \tag{7.28}$$

$$\textit{Blackman:} \quad w[n] = 0.42 + 0.5 \cos\left(\frac{2\pi n}{2M+1}\right) + 0.08 \cos\left(\frac{4\pi n}{2M+1}\right),$$
$$-M \leq n \leq M. \tag{7.29}$$

R7.16   The *Dolph-Chebyshev window* of length $2M + 1$ is an adjustable window defined by

$$w[n] = \frac{1}{2M+1}\left[\frac{1}{\gamma} + 2\sum_{k=1}^{M} T_k\left(\beta\,\cos\frac{k\pi}{2M+1}\right)\cos\frac{2nk\pi}{2M+1}\right],$$
$$-M \le n \le M, \qquad (7.30)$$

where

$$\gamma = \frac{\text{Aamplitude of side lobe}}{\text{Main lobe amplitude}}, \qquad (7.31)$$

$$\beta = \cosh\left(\frac{1}{2M}\cosh^{-1}\frac{1}{\gamma}\right), \qquad (7.32)$$

and $T_k(x)$ is the kth order Chebyshev polynomial defined by

$$T_k(x) = \begin{cases} \cos(k\,\cos^{-1}x), & \text{for } |x| \le 1, \\ \cosh(k\,\cosh^{-1}x), & \text{for } |x| > 1. \end{cases} \qquad (7.33)$$

R7.17   The most widely used adjustable window is the *Kaiser window* given by:

$$w[n] = \frac{I_0(\beta\sqrt{1-(n/M)^2})}{I_0(\beta)}, \qquad -M \le n \le M, \qquad (7.34)$$

where β is an adjustable parameter and $I_0(u)$ is the modified zeroth-order Bessel function, which can be expressed in a power series form:

$$I_0(u) = 1 + \sum_{r=1}^{\infty}\left[\frac{(u/2)^r}{r!}\right]^2. \qquad (7.35)$$

It can be seen that $I_0(u)$ is positive for all real values of u. In practice, it is sufficient to keep only the first 20 terms in the summation of Eq. (7.35) to arrive at a reasonably accurate value of $I_0(u)$. The parameter β controls the minimum attenuation α_s, that is, the ripple δ_s, in the stopband of the windowed filter response. Formulae for estimating β and the filter length $N = 2M + 1$, for specified α_s and transition bandwidth Δf, are given by

$$\beta = \begin{cases} 0.1102(\alpha_s - 8.7), & \text{for } \alpha_s > 50, \\ 0.5842(\alpha_s - 21)^{0.4} + 0.07886(\alpha_s - 21), & \text{for } 21 \le \alpha_s \le 50, \\ 0, & \text{for } \alpha_s < 21. \end{cases} \qquad (7.36)$$

and

$$N = \begin{cases} \frac{\alpha_s - 7.95}{14.36\,\Delta f} + 1, & \text{for } \alpha_s > 21, \\ \frac{0.9222}{\Delta f} + 1, & \text{for } \alpha_s \le 21. \end{cases} \qquad (7.37)$$

The Kaiser window provides no independent control over the passband ripple δ_p. However, in practice, δ_p is approximately equal to δ_s.

R7.18 The linear-phase FIR filter obtained by minimizing the peak absolute value of the weighted error ε given by

$$\varepsilon = \max_{0 \leq \omega \leq \pi} |\mathcal{E}(\omega)|, \tag{7.38}$$

is usually called the *optimal FIR filter* where the weighted error function $\mathcal{E}(\omega)$ defined by

$$\mathcal{E}(\omega) = P(\omega) \left[|H(e^{j\omega})| - D(\omega) \right], \tag{7.39}$$

exhibits an equiripple behavior in the frequency range of interest. The most widely used, highly efficient algorithm for designing the optimum linear-phase FIR filter is the *Parks-McClellan algorithm* [Par72].

7.3 MATLAB Commands Used

The MATLAB comands you will encounter in this exercise are as follows:

General Purpose Commands

```
disp        length
```

Operators and Special Characters

```
:        .        +        -        *        /        ;
%        .*       ./       >        =        ==
```

Language Constructs and Debugging

```
else        function        if
```

Elementary Matrices and Matrix Manipulation

```
fliplr      nargin      pi        :
```

Elementary Functions

```
abs        ceil        cos        log10        sin        sqrt
```

Data Analysis

```
min
```

Two-Dimensional Graphics

```
axis        grid        plot        title        xlabel
ylabel
```

Signal Processing Toolbox

blackman	butter	buttord	chebwin	cheb1ord
cheb2ord	cheby1	cheby2	ellip	ellipord
fir1	fir2	freqz	hanning	hamming
kaiser	remez	remezord		

For additional information on these commands, see the *MATLAB Reference Guide* [Mat94] and the *Signal Processing Toolbox User's Guide* [Mat96] or type help commandname in the Command window. A brief explanation of the MATLAB functions used here can be found in Appendix B.

7.4 IIR Filter Design

The most common method of IIR filter design is based on the bilinear transformation of a prototype analog transfer function. The analog transfer function is usually one of the following types: Butterworth, Type 1 Chebyshev, Type 2 Chebyshev, and elliptic transfer functions. The difference between these filter types can be explained by considering the analog lowpass filter. The Butterworth lowpass transfer function has a maximally flat magnitude response at dc, that is, $\Omega = 0$, and a monotonically decreasing magnitude response with increasing frequency. The Type 1 Chebyshev lowpass transfer function has an equiripple magnitude response in the passband and a monotonically decreasing magnitude response with increasing frequency outside the passband. The Type 2 Chebyshev lowpass transfer function has a monotonically decreasing magnitude response in the passband with increasing frequency and an equiripple magnitude response in the stopband. Finally, the elliptic lowpass transfer function has equiripple magnitude responses both in the passband and in the stopband.

Project 7.1 Estimation of Order of IIR Filter

The first step in the filter design process is to choose the type of filter approximation to be employed and then to estimate the order of the transfer function from the filter specifications. The MATLAB command for estimating the order of a Butterworth filter is

```
[N, Wn] = buttord(Wp, Ws, Rp, Rs),
```

where the input parameters are the normalized passband edge frequency Wp, the normalized stopband edge frequency Ws, the passband ripple Rp in dB, and the minimum stopband attenuation Rs in dB. Both Wp and Ws must be a number between 0 and 1 with the sampling frequency assumed to be 2 Hz. The output data are the lowest order N meeting the specifications and the normalized cutoff frequency Wn. If Rp = 3 dB, then Wn = Wp. buttord can also be used to estimate the order of a highpass, a bandpass, and a bandstop Butterworth filter. For a highpass filter design, Wp > Ws. For bandpass and bandstop filter designs, Wp and Ws are two-element vectors specifying both edge frequencies, with the lower edge frequency being the first element of the vector. In the latter cases, Wn is also a two-element vector.

For estimating the order of a Type 1 Chebyshev filter, the MATLAB command is

```
[N, Wn] = cheb1ord(Wp, Ws, Rp, Rs)
```

and for designing a Type 2 Chebyshev filter, the MATLAB command for estimating the order is

```
[N, Wn] = cheb2ord(Wp, Ws, Rp, Rs).
```

Finally, in the case of an elliptic filter design, the command is

```
[N, Wn] = ellipord(Wp, Ws, Rp, Rs).
```

As before, Wp and Ws are the passband and stopband edge frequencies with values between 0 and 1. Likewise, Rp and Rs are the passband ripple and the minimum stopband attenuation in dB. N contains the estimated lowest order and Wn is the cutoff frequency. It should be noted that for bandpass and bandstop filter designs, the actual order of the transfer function obtained using the appropriate filter design command is 2N.

Questions:

Q7.1 Using MATLAB determine the lowest order of a digital IIR lowpass filter of all four types. The specifications are as follows: sampling rate of 40 kHz, passband edge frequency of 4 kHz, stopband edge frequency of 8 kHz, passband ripple of 0.5 dB, and a minimum stopband attenuation of 40 dB. Comment on your results.

Q7.2 Using MATLAB determine the lowest order of a digital IIR highpass filter of all four types. The specifications are as follows: sampling rate of 3,500 Hz, passband edge frequency of 1,050 Hz, stopband edge frequency of 600 Hz, passband ripple of 1 dB, and a minimum stopband attenuation of 50 dB. Comment on your results.

Q7.3 Using MATLAB determine the lowest order of a digital IIR bandpass filter of all four types. The specifications are as follows: sampling rate of 7 kHz, passband edge frequencies at 1.4 kHz and 2.1 kHz, stopband edge frequencies at 1.05 kHz and 2.45 kHz, passband ripple of 0.4 dB, and a minimum stopband attenuation of 50 dB. Comment on your results.

Q7.4 Using MATLAB determine the lowest order of a digital IIR bandstop filter of all four types. The specifications are as follows: sampling rate of 12 kHz, passband edge frequencies at 2.1 kHz and 4.5 kHz, stopband edge frequencies at 2.7 kHz and 3.9 kHz, passband ripple of 0.6 dB, and a minimum stopband attenuation of 45 dB. Comment on your results.

Project 7.2   IIR Filter Design

After the filter type has been selected and its order estimated, the next step is to determine the transfer function of the filter. To this end MATLAB provides functions for all four types of filters. For designing Butterworth digital lowpass or bandpass filters, the command is

```
[num,den] = butter(N,Wn)
```

where the input parameters N and Wn are determined through the use of the function buttord, and the output is the vectors num and den containing, respectively, the coefficients of the numerator and denominator polynomials of the transfer function in ascending powers of z^{-1}. If Wn is a scalar, butter returns a lowpass transfer function of order N, and if Wn is a two-element vector, it returns a bandpass transfer function of order 2N. For designing a Butterworth digital highpass filter of order N, the command is

```
[num,den] = butter(N,Wn,'high')
```

whereas, the command

```
[num,den] = butter(N,Wn,'stop')
```

returns the transfer function of a Butterworth bandstop filter of order 2N provided Wn is a two-element vector. For designing a Type 1 Chebyshev digital filter, the commands are

```
[num,den] = cheby1(N,Rp,Wn)
[num,den] = cheby1(N,Rp,Wn,'filtertype')
```

For designing a Type 2 Chebyshev digital filter, the commands are

```
[num,den] = cheby2(N,Rs,Wn)
[num,den] = cheby2(N,Rs,Wn,'filtertype')
```

Finally, for designing an elliptic digital filter, the commands are

```
[num,den] = ellip(N,Rp,Rs,Wn)
[num,den] = ellip(N,Rp,Rs,Wn,'filtertype')
```

A lowpass transfer function of order N is returned in each case if Wn is a scalar, and a bandpass transfer function of order 2N is returned if Wn is a two-element vector. In each of the above commands, filtertype is high for designing a highpass filter with Wn being a scalar, and filtertype is stop for designing a bandstop filter with Wn being a two-element vector.

Program P7_1 illustrates the design of a Butterworth bandstop filter.

```
% Program P7_1
% Design of a Butterworth Bandstop Digital Filter
Ws = [0.4 0.6]; Wp = [0.3 0.7]; Rp = 0.4; Rs = 50;
% Estimate the Filter Order
[N1, Wn1] = buttord(Wp, Ws, Rp, Rs);
```

```
% Design the Filter
[num,den] = butter(N1,Wn1,'stop');
% Display the transfer function
disp('Numerator coefficients are ');disp(num);
disp('Denominator coefficients are ');disp(den);
% Compute the gain response
[g,w] = gain(num,den);
% Plot the gain response
plot(w/pi,g);grid
axis([0 1 -60 5]);
xlabel('\omega /\pi');
ylabel('Gain, dB');
title('Gain Response of a Butterworth Bandstop Filter');
```

Questions:

Q7.5 Design the Butterworth bandstop filter by running Program P7_1. Write down the exact expression for the transfer function generated. What are the filter specifications? Does your design meet the specifications? Using MATLAB, compute and plot the filter's unwrapped phase response and the group delay response.

Q7.6 Modify Program P7_1 to design a Type 1 Chebyshev lowpass filter meeting the specifications given in Question Q7.1. Write down the exact expression for the transfer function generated. Does your design meet the specifications? Using MATLAB, compute and plot the filter's unwrapped phase response and the group delay response.

Q7.7 Modify Program P7_1 to design a Type 2 Chebyshev highpass filter meeting the specifications given in Question Q7.2. Write down the exact expression for the transfer function generated. Does your design meet the specifications? Using MATLAB, compute and plot the filter's unwrapped phase response and the group delay response.

Q7.8 Modify Program P7_1 to design an elliptic bandpass filter meeting the specifications given in Question Q7.3. Write down the exact expression for the transfer function generated. Does your design meet the specifications? Using MATLAB, compute and plot the filter's unwrapped phase response and the group delay response.

7.5 FIR Filter Design

Conceptually the simplest approach to FIR filter design is to simply truncate to a finite number of terms the doubly infinite-length impulse response coefficients obtained by computing the inverse discrete-time Fourier transform of the desired ideal frequency response. However, a simple truncation results in an oscillatory behavior in the respective magnitude response of the FIR filter, which is more commonly referred to as the *Gibbs phenomenon.*

The Gibb's phenomenon can be reduced by windowing the doubly infinite-length impulse response coefficients by an appropriate finite-length window function. The functions `fir1` and `fir2` can be employed to design windowed FIR digital filters in MATLAB. Both functions yield a linear-phase design.

The function `fir1` can be used to design conventional lowpass, highpass, bandpass, and bandstop linear-phase FIR filters. The command

```
b = fir1(N,Wn)
```

returns in vector `b` the impulse response coefficients, arranged in ascending powers of z^{-1}, of a lowpass or a bandpass filter of order `N` for an assumed sampling frequency of 2 Hz. For lowpass design, the normalized cutoff frequency is specified by a scalar `Wn`, a number between 0 and 1. For bandpass design, `Wn` is a two-element vector [`Wn1`, `Wn2`] containing the specified passband edges where 0 < `Wn1` < `Wn2` < 1. The command

```
b = fir1(N,Wn,'high')
```

with `N` an even integer, is used for designing a highpass filter. The command

```
b = fir1(N,Wn,'stop')
```

with `Wn` a two-element vector, is employed for designing a bandstop FIR filter. If none is specified, the *Hamming window* is employed as a default. The command

```
b = fir1(N, Wn, taper)
```

makes use of the specified window coefficients of length N+1 in the vector `taper`. However, the window coefficients must be generated a priori using an appropriate MATLAB function such as `blackman`, `hamming`, `hanning`, `chebwin`, or `kaiser`. The commands to use are of the following forms:

```
taper = blackman(N)    taper = hamming(N)    taper = hanning(N)
taper = chebwin(N)     taper = kaiser(N, beta)
```

The function `fir2` can be used to design linear-phase FIR filters with arbitarily shaped magnitude responses. In its basic form, the command is

```
b = fir2(N, fpts, mval)
```

which returns in the vector `b` of length N+1 the impulse response coefficients, arranged in ascending powers of z^{-1}. `fpts` is the vector of specified frequency points, arranged in an increasing order, in the range 0 to 1 with the first frequency point being 0 and the last frequency point being 1. As before, the sampling frequency is assumed to be 2 Hz. `mval` is a vector of specified magnitude values at the specified frequency points and therefore must also be of the same length as `fpts`. The Hamming window is used as a default. To make use of other windows, the command to use is

```
b = fir2(N, fpts, mval,taper)
```

where the vector `taper` contains the specified window coefficients.

A more widely used linear-phase FIR filter design is based on the Parks-McClellan algorithm , which results in an optimal FIR filter with an equiripple weighted error $\mathcal{E}(\omega)$ defined in Eq. (7.39). It makes use of the Remez optimization algorithm and is available in MATLAB as the function `remez` . This function can be used to design any type of single-band or multiband filter, the differentiator, and the Hilbert transformer. In its basic form, the command

```
b = remez(N,fpts,mval)
```

returns a vector b of length N+1 containing the impulse response coefficients of the desired FIR filter in ascending powers of z^{-1}. `fpts` is the vector of specified frequency points, arranged in increasing order, in the range 0 to 1 with the first frequency point being 0 and the last frequency point being 1. As before, the sampling frequency is assumed to be 2 Hz. The desired magnitudes of the FIR filter frequency response at the specified band edges are given by the vector `mval`, with the elements given in equal-valued pairs. The desired magnitudes between two specified consecutive frequency points `f(k)` and `f(k+1)` are determined according to the following rules. For k odd, the magnitude is a line segment joining the points {`mval(k)`, `fpts(k)`} and {`mval(k+1)`, `fpts(k+1)`}, whereas, for k even, it is unspecified with the frequency range [`fpts(k)`, `fpts(k+1)`] being a *transition* or *"don't care"* region. The vectors `fpts` and `mval` must be of the same length with the length being even. Figure 7.4 illustrates the relationship between the vectors `fpts` and `mval` given by

```
fpts = [0 0.2 0.4 0.7 0.8 1.0]
mval = [0.5 0.5 1.0 1.0 0.3 0.3]
```

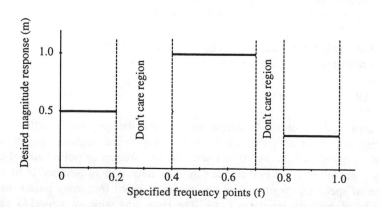

Figure 7.4   Illustration of relationship between vectors `fpts` and `mval`.

The desired magnitude responses in the passband(s) and the stopband(s) can be weighted by an additional vector `wgts` included as the argument of the function `remez`. The function can be used to design equiripple Types 1, 2, 3, and 4 linear-phase FIR filters. Types 1 and 2 are the default designs for order N even and odd, respectively. Types 3 (N even) and 4 (N odd) are used for specialized filter designs, the Hilbert transformer and the differentiator. To design these two types of FIR filters the flags `hilbert` and `differentiator` are used for `ftype` in the last two versions of `remez`. The command

```
b = remez(N,fpts,mval,wgts)
```

is used to design an FIR filter weighted in each band by the elements of the weight vector `wgts` whose length is thus half that of `fpts`. The elements of the vector `wgts` can be determined from the specified passband and stopband ripples by dividing the maximum ripple value by the ripple values. To design a Hilbert transformer or a differentiator, use the forms

```
remez(N,fpts,mval,ftype)
remez(N,fpts,mval,wgts,ftype)
```

where `ftype` is the string `hilbert` or `differentiator`. In the case of a Hilbert transformer design, the smallest element in `fpts` should not be a 0.

The order N of the FIR filter to meet the given specifications can be estimated using either Kaiser's formula of Eq. (7.7) or Hermann's formula of Eq. (7.8). The MATLAB function `kaiord` given below implements Kaiser's formula:

```
function N = kaiord(Fp, Fs, dp, ds, FT)
% Computation of the length of a linear-phase
% FIR multiband filter using Kaiser's formula
% dp is the passband ripple
% ds is the stopband ripple
% Fp is the passband edge in Hz
% Fs is the stopband edge in Hz
% FT is the sampling frequency in Hz.
% If none specified default value is 2
% N is the estimated FIR filter order
if nargin == 4,
    FT = 2;
end
if length(Fp) > 1,
    TBW = min(abs(Fp(1) - Fs(1)), abs(Fp(2) - Fs(2)));
    else
    TBW = abs(Fp - Fs);
end
num = -20*log10(sqrt(dp*ds)) - 13;
den = 14.6*TBW/FT;
N = ceil(num/den);
```

The function `kaiserord` in the Signal Processing Toolbox can also be used for estimating the filter order using Kaiser's formula. It can be used in one of the following forms:

```
[N, Wn, beta, ftype] = kaiserord(fedge, aval, dev)
[N, Wn, beta, ftype] = kaiserord(fedge, aval, dev, FT)
c = kaiserord(fedge, aval, dev, FT, 'cell')
```

where FT is the sampling frequency in Hz whose default value is 2 Hz if not specified; `fedge` is a vector of bandedge frequencies in Hz, in increasing order between 0 and FT/2; and `aval` is a vector specifying the desired values of the magnitude response at the specified bandedges given by `fedge`. The length of `fedge` is 2 less than twice the length of `aval` and therefore must be even. `dev` is a vector of maximum deviations or ripples in dB allowable for each band. If the deviations specified are unequal, the smallest one is used for all bands. The output data are in the desired format for use in `fir1`, with normalized bandedges Wn and the parameter `beta` used for computing the window coefficients as given in Eq. (7.36). The string `ftype` specifies the filter type for `fir1`. It is high for highpass filter design, and `stop` for bandstop filter design. The last form of `kaiserord` specifies a cell array whose elements are parameters to `fir1`.

The MATLAB function `remezord` implements the formula of Eq. (7.8). It can be used in one of the following forms:

```
[N,fts,mval,wgts] = remezord(fedge,aval,dev)
[N,fts,mval,wgts] = remezord(fedge,aval,dev,FT)
```

where FT is the sampling frequency in Hz whose default value is 2 Hz if not specified, `fedge` is a vector of bandedge frequencies in Hz, in increasing order between 0 and FT/2; and `aval` is a vector specifying the desired values of the magnitude response at the specified bandedges given by `fedge`. The length of `fedge` is 2 less than twice the length of `aval` and therefore must be even. `dev` is a vector of maximum deviations or ripples in dB allowable for each band. A third form of `remezord` is given by

```
c = remezord(fedge,aval,dev,FT, 'cell')
```

and specifies a cell array whose elements are the parameters to `remez`.

In some cases, the order N determined using either method may not result in an FIR filter meeting the original specifications. If it does not, the order should either be increased or decreased by 1 gradually until the specifications are met. Moreover, the order estimates may be highly inaccurate for very narrowband or very wideband FIR filters.

Project 7.3   Gibb's Phenomenon

The occurrence of Gibb's phenomenon can be illustrated by considering the design of an FIR filter obtained by truncating the impulse response of the ideal filters given by Eqs.

(7.16), (7.18) – (7.20), (7.22), (7.24) and (7.26), and then computing their frequency responses. The truncated impulse response coefficients of a lowpass filter can be generated in MATLAB using the function sinc, which can also be used with simple modifications to generate the truncated impulse response coefficients of a highpass, bandpass, or bandstop filter.

Questions:

Q7.9 Using the function sinc write a MATLAB program to generate the impulse response coefficients of four zero-phase lowpass filters with cutoffs at $\omega_c = 0.4\pi$ and of lengths 81, 61, 41, and 21, respectively, and then compute and plot their magnitude responses. Use the colon ":" operator to extract the impulse response coefficients of the shorter length filters from that of the length-81 filter. Examine the oscillatory behavior of the frequency responses of each filter on both sides of the cutoff frequency. What is the relation between the number of ripples and the length of the filter? What is the relation between the heights of the largest ripples and the length of the filter? How would you modify the above program to generate the impulse response coefficients of a zero-phase lowpass filter of even lengths?

Q7.10 Using the function sinc write a MATLAB program to generate the impulse response coefficients of a zero-phase length-45 highpass filter with a cutoff at $\omega_c = 0.4\pi$ and then compute and plot its magnitude response. Examine the oscillatory behavior of the frequency responses of each filter on both sides of the cutoff frequency. How would you modify the above program to generate the impulse response coefficients of a zero-phase highpass filter of even length?

Q7.11 Write a MATLAB program to generate the impulse response coefficients of four zero-phase differentiators of lengths 81, 61, 41, and 21, respectively, and then compute and plot their magnitude responses. The following code fragments show how to generate a differentiator of length 2M+1.

```
n = 1:M;
b = cos(pi*n)./n;
num = [-fliplr(b) 0 b];
```

Examine the oscillatory behavior of the frequency response of the differentiator for each case. What is the relation between the number of ripples and the length of the differentiator? What is the relation between the heights of the largest ripples and the length of the filter?

Q7.12 Write a MATLAB program to generate the impulse response coefficients of four discrete-time Hilbert transformers of lengths 81, 61, 41, and 21, respectively, and then compute and plot their magnitude responses. The following code fragments show how to generate a Hilbert transformer of length 2M+1.

```
n = 1:M;
c = sin(pi*n/2);
```

```
b = 2*(c.*c)./(pi*n);
num = [-fliplr(b) 0 b];
```

Examine the oscillatory behavior of the frequency responses of the Hilbert transformer for each case. What is the relation between the number of ripples and the length of the Hilbert transformer? What is the relation between the heights of the largest ripples and the length of the filter?

Project 7.4   Estimation of Order of FIR Filter

Questions:

Q7.13   Using the function `kaiord`, estimate the order of a linear-phase lowpass FIR filter with the following specifications: passband edge = 2 kHz, stopband edge = 2.5 kHz, passband ripple $\delta_p = 0.005$, stopband ripple $\delta_s = 0.005$, and sampling rate of 10 kHz. What are the purposes of the commands `ceil` and `nargin` in the function `kaiord`?

Q7.14   Repeat Question Q7.13 for the following cases: (a) sampling rate of 20 kHz, (b) $\delta_p = 0.002$ and $\delta_s = 0.002$, and (c) stopband edge = 2.3 kHz. Compare the filter length obtained in each case with that obtained in Question Q7.13. Comment on the effect of the sampling rate, ripples, and the transition bandwidth on the filter order.

Q7.15   Repeat Question Q7.13 using the function `kaiserord`. Compare the value of the filter order obtained with that obtained in Question Q7.13.

Q7.16   Repeat Question Q7.13 using the function `remezord`. Compare the value of the filter order obtained with those obtained in Questions Q7.13 and Q7.15.

Q7.17   Using the function `kaiord`, estimate the order of a linear-phase bandpass FIR filter with the following specifications: passband edges = 1.8 and 3.6 kHz, stopband edges = 1.2 and 4.2 kHz, passband ripple $\delta_p = 0.1$, stopband ripple $\delta_s = 0.02$, and sampling rate of 12 kHz.

Q7.18   Repeat Question Q7.17 using the function `kaiserord`. Compare the value of the filter order obtained with that obtained in Question Q7.17.

Q7.19   Repeat Question Q7.17 using the function `remezord`. Compare the value of the filter order obtained with that obtained in Questions Q7.17 and Q7.18.

Project 7.5   FIR Filter Design

Questions:

Q7.20   Using the function `fir1`, design a linear-phase FIR lowpass filter meeting the specifications given in Question Q7.13 and plot its gain and phase responses. Use the order estimated using Kaiser's formula in Question Q7.13. Show the filter coefficients in a tabular form. Does your design meet the specifications? If it does not, adjust the filter order until the design meets the specifications. What is the order of the filter meeting the specifications?

Q7.21 Repeat Question Q7.20 using each of the following windows: Hanning, Blackman, and Dolph-Chebyshev windows.

Q7.22 Repeat Question Q7.20 using the function `remez`.

Q7.23 Design an FIR lowpass filter using a Kaiser window. The filter specifications are: $\omega_p = 0.3$, $\omega_s = 0.4$, and $A_s = 50$ dB. Note that the function `kaiser` requires the values of the parameter β and the order N which must first be calculated using Eqs. (7.36) and (7.37), respectively. Does your design meet the specifications?

Q7.24 Repeat Question Q7.23 using the functions `kaiserord` and `fir1`.

Q7.25 Using `fir2` design an FIR filter of order 95 with three different constant magnitude levels: 0.4 in the frequency range 0 to 0.25, 1.0 in the frequency range 0.3 to 0.45, and 0.8 in the frequency range 0.5 to 1.0. Plot the magnitude response of the filter designed. Does your design meet the specifications?

Q7.26 Using `remez`, design the FIR bandpass filter with specifications given in Question Q7.17 and of order obtained using `kaiserord`. The vector `wgts` needed in the function `remez` is given by

$$\texttt{wgts} = \max(\delta_p, \delta_s)[1/\delta_p, 1/\delta_s]$$

Does your design meet the specifications? If it does not, adjust the filter order until the design meets the specifications. What is the order of the filter meeting the specifications?

Q7.27 Using `remez`, design an FIR bandpass filter with the following specifications: passband edges = 1.8 and 3.0 kHz, stopband edges = 1.5 and 4.2 kHz, passband ripple $\delta_p = 0.1$, stopband ripple $\delta_s = 0.02$, and sampling rate of 12 kHz. Estimate the filter order using `kaiserord`. Is your design an optimal FIR filter? Does your design meet the specifications? If it does not, does increasing the filter order help in meeting the specifications? Are the specifications met by a filter with a lower order than that obtained using `kaiserord`? In the case of unequal transition bands, the filter designed using `remez` may exhibit unsatisfactory behavior in the larger transition band in its gain response. One way to improve the behavior is to reduce the transition band by moving the stopband edge until the design meets the specifications with smooth roll-off in the transition bands. Try this approach and determine the new specifications, with passband edges remaining fixed, that provide smooth roll-off in the transition bands.

7.6 Background Reading

[1] A. Antoniou. *Digital Filters: Analysis, Design, and Applications*. McGraw-Hill, New York NY, second edition, 1993. Chs. 7, 9.

[2] E. Cunningham. *Digital Filtering: An Introduction*. Houghton-Mifflin, Boston MA, 1992. Secs. 4.2, 4.3, 4.5, 4.6, 5.2 – 5.5.

[3] D.J. DeFatta, J.G. Lucas, and W.S. Hodgkiss. *Digital Signal Processing: A System Design Approach.* Wiley, New York NY, 1988. Chs. 4, 5.

[4] L.B. Jackson. *Digital Filters and Signal Processing.* Kluwer, Boston, MA, third edition, 1996. Chs. 8, 9.

[5] R. Kuc. *Introduction to Digital Signal Processing.* McGraw-Hill, New York NY, 1988. Chs. 8, 9.

[6] L.C. Ludeman. *Fundamentals of Digital Signal Processing.* Harper & Row, New York NY. 1986. Ch. 4.

[7] S.K. Mitra. *Digital Signal Processing: A Computer-Based Approach.* McGraw-Hill, New York NY, 1998. Ch. 7.

[8] A.V. Oppenheim and R.W. Schafer. *Discrete-Time Signal Processing.* Prentice-Hall, Englewood Cliffs NJ, 1989. Ch. 7.

[9] S.J. Orfanidis. *Introduction to Signal Processing.* Prentice-Hall, Englewood Cliffs NJ, 1996. Ch. 2 and Chs. 10, 11.

[10] B. Porat. *A Course in Digital Signal Procesing.* Wiley, New York NY, 1996. Secs. 9.2–9.6, 10.9.

[11] J.G. Proakis and D.G. Manolakis. *Digital Signal Processing: Principles, Algorithms, and Applications*, Prentice-Hall, Englewood Cliffs NJ, third edition, 1996, Ch. 8.

[12] R.A. Roberts and C.T. Mullis. *Digital Signal Processing.* Addison-Wesley, Reading MA, 1987. Ch. 6.

Digital Filter Implementation

8.1 Introduction

After the digital transfer function $G(z)$, obtained by approximating the given frequency response specifications, has been realized in a suitable filter form, the structure is next implemented in either hardware or software form. In this laboratory exercise you will investigate the software implementation of an IIR or FIR digital filter structure using MATLAB.

8.2 Background Review

R8.1 A simple technique to verify the structure is based on the convolution sum relation between the transfer function coefficients and the filter's impulse response samples [Mit77b]. Let $H(z)$ be an Nth order causal IIR transfer function given by

$$H(z) = \frac{p_0 + p_1\, z^{-1} + p_2\, z^{-2} + \ldots + p_N\, z^{-N}}{1 + d_1\, z^{-1} + d_2\, z^{-2} + \ldots + d_N\, z^{-N}}. \tag{8.1}$$

If $\{h[n]\}$ denotes its unit sample response, then

$$H(z) = \sum_{n=0}^{\infty} h[n]\, z^{-n}. \tag{8.2}$$

From Eqs. (8.1) and (8.2) it follows that

$$p_n = h[n] \circledast d_n, \qquad n = 0, 1, 2, \ldots. \tag{8.3}$$

The first $2N + 1$ equations from Eq. (8.3) can be expressed in matrix form as

$$\mathbf{p} = \mathbf{H}_1 \begin{bmatrix} 1 \\ \mathbf{d} \end{bmatrix}, \tag{8.4}$$

$$\mathbf{0} = [\mathbf{h} \quad \mathbf{H}_2] \begin{bmatrix} 1 \\ \mathbf{d} \end{bmatrix}, \tag{8.5}$$

where

$$\mathbf{p} = \begin{bmatrix} p_0 \\ p_1 \\ \vdots \\ p_N \end{bmatrix}, \quad \mathbf{d} = \begin{bmatrix} d_1 \\ d_2 \\ \vdots \\ d_N \end{bmatrix}, \quad \mathbf{0} = \begin{bmatrix} 0 \\ \vdots \\ 0 \end{bmatrix}, \tag{8.6}$$

$$\mathbf{H}_1 = \begin{bmatrix} h[0] & 0 & \cdots & 0 \\ h[1] & h[0] & \cdots & 0 \\ \vdots & \vdots & \ddots & \vdots \\ h[N] & h[N-1] & \cdots & h[0] \end{bmatrix}, \tag{8.7}$$

$$\mathbf{h} = \begin{bmatrix} h[N+1] \\ \vdots \\ h[2N] \end{bmatrix}, \qquad \mathbf{H}_2 = \begin{bmatrix} h[N] & \cdots & h[1] \\ \vdots & \ddots & \vdots \\ h[2N-1] & \cdots & h[N] \end{bmatrix}, \tag{8.8}$$

Solving Eq. (8.5) the vector $\mathbf{d}$ composed of the denominator coefficients

$$\mathbf{d} = -\mathbf{H}_2^{-1}\mathbf{h}, \tag{8.9}$$

is first obtained. Substituting Eq. (8.9) in Eq. (8.4) the vector $\mathbf{p}$ containing the numerator coefficients

$$\mathbf{p} = \mathbf{H}_1 \begin{bmatrix} 1 \\ -\mathbf{H}_2^{-1}\mathbf{h} \end{bmatrix}. \tag{8.10}$$

is next determined.

R8.2   The basis for the design of tunable digital filters is the spectral transformation of the complex variable z by which a given digital filter realization with a specified cutoff frequency can be tuned to arrive at another realization with a different cutoff frequency by replacing z with $F(z)$. Thus, if $G_{old}(z)$ is the transfer function of the original realization, the transfer function of the new structure is $G_{new}(z)$ where

$$G_{new}(z) = G_{old}(z)|_{z^{-1}=F^{-1}(z)}, \tag{8.11}$$

where $F^{-1}(z)$ is an appropriately chosen stable allpass function with coefficients that are the tuning parameters [Con70]. One straightforward way to implement this transformation would be to replace each delay block in the realization of $G_{old}(z)$ with an allpass structure realizing $F^{-1}(z)$. However, such an approach leads, in general, to a structure realizing $G_{new}(z)$ with delay-free loops that cannot be implemented.

A very simple practical modification of the above approach applied to a parallel allpass realization does not result in a structure with delay-free loops [Mit90]. A bounded real transfer function $G(z)$ with a symmetric numerator and having a power complementary transfer function $H(z)$ with an antisymmetric numerator can be realized in the form

$$G(z) = \frac{1}{2}\{A_0(z) + A_1(z)\}, \tag{8.12}$$

where $A_0(z)$ and $A_1(z)$ are stable allpass filters. The above conditions on $G(z)$ are satisfied by all odd-order lowpass Butterworth, Chebyshev, and elliptic transfer functions.

The allpass filters $A_0(z)$ and $A_1(z)$ are realized as a cascade of first-order and second-order sections of the form depicted in Figures 6.9–6.11. These structures use only one multiplier and one delay for the realization of a first-order allpass function, and two multipliers and two delays for the realization of a second-order allpass function.

For the tuning of the cutoff frequency of a lowpass IIR filter realized by a parallel allpass structure, the lowpass-to-lowpass transformation given by [Con70]

$$z^{-1} \rightarrow F(z^{-1}) = \frac{z^{-1} - \alpha}{1 - \alpha\, z^{-1}}, \tag{8.13}$$

is employed where the parameter α is related to the old and new cutoff frequencies, ω_c and $\hat{\omega}_c$, respectively, through

$$\alpha = \frac{\sin\left[(\omega_c - \hat{\omega}_c)/2\right]}{\sin\left[(\omega_c + \hat{\omega}_c)/2\right]}. \tag{8.14}$$

Substitution of the transformation of Eq. (8.13) in a Type 1 first-order allpass transfer function[1]

$$a_1(z) = \frac{d_1 + z^{-1}}{1 + d_1\, z^{-1}}, \tag{8.15}$$

results in a transformed first-order allpass transfer function given by

$$\hat{a}_1(z) = a_1(z)\big|_{z^{-1} = \frac{z^{-1} - \alpha}{1 - \alpha z^{-1}}} \cong \frac{\left[d_1 + \alpha(d_1^2 - 1)\right] + z^{-1}}{1 + \left[d_1 + \alpha(d_1^2 - 1)\right] z^{-1}}, \tag{8.16}$$

assuming α to be small. $\hat{a}_1(z)$ is a Type 1 first-order allpass transfer function with a coefficient that is now a linear function of α. In the case of a Type 3 second-order allpass transfer function[2]

$$a_2(z) = \frac{d_2 + d_1 z^{-1} + z^{-2}}{1 + d_1 z^{-1} + d_2 z^{-2}}, \tag{8.17}$$

the transformation of Eq. (8.13) yields

$$\hat{a}_2(z) = a_2(z)\big|_{z^{-1} = \frac{z^{-1} - \alpha}{1 - \alpha z^{-1}}} = \frac{d_2 + d_1\left(\frac{z^{-1} - \alpha}{1 - \alpha z^{-1}}\right) + \left(\frac{z^{-1} - \alpha}{1 - \alpha z^{-1}}\right)^2}{1 + d_1\left(\frac{z^{-1} - \alpha}{1 - \alpha z^{-1}}\right) + d_2\left(\frac{z^{-1} - \alpha}{1 - \alpha z^{-1}}\right)^2}$$

$$\cong \frac{\left[d_2 + \alpha\, d_1(d_2 - 1)\right] + \left[d_1 - 2\,\alpha(1 + d_2) + \alpha\, d_1^2\right] z^{-1} + z^{-2}}{1 + \left[d_1 - 2\,\alpha(1 + d_2) + \alpha\, d_1^2\right] z^{-1} + \left[d_2 + \alpha\, d_1(d_2 - 1)\right] z^{-2}}, \tag{8.18}$$

which is seen to be a Type 3 second-order allpass transfer function with coefficients that are linear functions of α.

By applying a lowpass-to-bandpass transformation [Con70]

$$z^{-1} \rightarrow F(z^{-1}) = -z^{-1} \frac{z^{-1} + \beta}{1 + \beta\, z^{-1}}, \tag{8.19}$$

to a tunable lowpass IIR filter, we can design a tunable bandpass filter whose center frequency ω_o is tuned by adjusting the parameter $\beta = \cos \omega_o$ and whose bandwidth is tuned by changing α. Unlike the lowpass-to-lowpass transformation of Eq. (8.13), the transformation can be directly implemented on the structure realizing the tunable lowpass filter by replacing each delay with an allpass structure cascaded with a delay.

---

[1]See R6.13.
[2]see R6.15.

R8.3   A very simple approach to the design of a tunable FIR filter is based on the windowed Fourier series approach [Jar88]. For an ideal FIR lowpass filter $H_d(z)$ with a zero-phase response

$$H_d(e^{j\omega}) = \begin{cases} 1, & \text{for } 0 \le \omega \le \omega_c, \\ 0, & \text{for } \omega_c \le \omega \le \pi, \end{cases} \tag{8.20}$$

the impulse response coefficients are given by

$$h_d[n] = \frac{\sin(\omega_c n)}{\pi n}, \qquad 0 \le |n| < \infty. \tag{8.21}$$

The above expression is then truncated to arrive at the coefficients of a realizable approximation given by

$$h_{LP}[n] = \begin{cases} c[n]\,\omega_c, & \text{for } n = 0, \\ c[n]\,\sin(\omega_c n), & \text{for } 1 \le |n| \le N, \\ 0, & \text{otherwise,} \end{cases} \tag{8.22}$$

where ω_c is the 6-dB cutoff frequency and

$$c[n] = \begin{cases} 1/\pi, & \text{for } n = 0, \\ 1/\pi n, & \text{for } 1 \le |n| \le N. \end{cases} \tag{8.23}$$

It follows from the above that once an FIR lowpass filter has been designed for a given cutoff frequency, it can be tuned simply by changing ω_c and recomputing the filter coefficients according to the above expression. It can be shown that Eq. (8.22) can also be used to design a tunable FIR lowpass filter by equating the coefficients of a prototype filter developed using any of the FIR filter design methods with those of Eq. (8.22) and solving for $c[n]$. Thus, if $h_{LP}[n]$ denotes the coefficients of the prototype lowpass filter designed for a cutoff frequency ω_c, from Eq. (8.22) the constants $c[n]$ are given by

$$c[0] = \frac{h_{LP}[0]}{\omega_c},$$

$$c[n] = \frac{h_{LP}[n]}{\sin(\omega_c n)}, \qquad 1 \le |n| \le N. \tag{8.24}$$

Then, the coefficients $\hat{h}_{LP}[n]$ of the transformed FIR filter with a cutoff frequency $\hat{\omega}_c$ are given by

$$\hat{h}_{LP}[0] = c[0]\,\hat{\omega}_c = \left(\frac{\hat{\omega}_c}{\omega_c}\right) h_{LP}[0],$$

$$\hat{h}_{LP}[n] = c[n]\,\sin(\hat{\omega}_c n) = \left(\frac{\sin(\hat{\omega}_c n)}{\sin(\omega_c n)}\right) h_{LP}[n], \qquad 1 \le |n| \le N. \tag{8.25}$$

This tuning procedure works well for filters with equal passband and stopband ripples. The prototype filter should be designed such that its coefficients have values not too close to zero.

R8.4 The N-point discrete Fourier transform (DFT) $X[k], 0 \le k \le N - 1$, of a length-N sequence $x[n], 0 \le n \le N - 1$, is given by the N samples of its z-transform $X(z)$ evaluated on the unit circle at N equally spaced points $z = e^{j2\pi k/N}$, $0 \le k \le N - 1$:

$$X[k] = X(z)|_{z=e^{j2\pi k/N}} = \sum_{n=0}^{N-1} x[n]e^{-j2\pi kn}, \qquad 0 \le k \le N - 1. \tag{8.26}$$

A direct computation of the N-point DFT using Eq. (8.26) therefore requires, for a complex-valued length-N sequence, N^2 complex multiplications and $N(N - 1) \approx N^2$ complex additions.

If N is even, $X(z)$ can be expressed in a two-band polyphase form as (see R10.9):

$$X(z) = X_0(z^2) + z^{-1}X_1(z^2), \tag{8.27}$$

where $X_0(z)$ is the z-transform of the length-$(N/2)$ subsequence formed from the even-indexed samples and $X_1(z)$ is the z-transform of the length-$(N/2)$ subsequence formed from the odd-indexed samples of $x[n]$, that is,

$$X_0(z) = \sum_{n=0}^{\frac{N}{2}-1} x_0[n]z^{-n} = \sum_{n=0}^{\frac{N}{2}-1} x[2n]z^{-n}, \tag{8.28}$$

$$X_1(z) = \sum_{n=0}^{\frac{N}{2}-1} x_1[n]z^{-n} = \sum_{n=0}^{\frac{N}{2}-1} x[2n + 1]z^{-n}, \tag{8.29}$$

Hence, the N-point DFT $X[k]$ can be computed using

$$X[k] = X_0[\langle k \rangle_{N/2}] + W_N^k X_1[\langle k \rangle_{N/2}], \qquad 0 \le k \le N - 1, \tag{8.30}$$

where $X_0[k]$ and $X_1[k]$ are the $\frac{N}{2}$-point DFTs of the $\frac{N}{2}$-length sequences, $x_0[n]$ and $x_1[n]$, respectively. An implementation of the N-point DFT using Eq. (8.30) now requires $N + \frac{N^2}{2}$ complex multiplications and approximately $N + \frac{N^2}{2}$ complex additions. This process can be continued if N is a power of 2 until the N-point DFT computation reduces to a weighted combination of length-2 DFTs. This is the basic idea behind the *decimation-in-time fast Fourier transform (FFT)* algorithm. It can be shown that in the general case, the total number of complex multiplications and complex additions is $N(\log_2 N)$. Further reduction in computational complexity is possible by exploiting the symmetry property of W_N.

R8.5 The sine of a number x can be approximated using the expansion [Abr72]

$$\sin(x) \cong x - 0.166667\,x^3 + 0.008333\,x^5 - 0.0001984\,x^7 + 0.0000027\,x^9, \tag{8.31}$$

where the argument x is in radians, and its range is restricted to the first quadrant, that is, from 0 to $\pi/2$. If x is outside this range, its sine can be computed by making use of the identities: $\sin(-x) = -\sin(x)$, and $\sin(\frac{\pi}{2} + x) = \sin(\frac{\pi}{2} - x)$.

R8.6   The arctangent of a number x, where $-1 \leq x \leq 1$, can be computed by [Abr72]

$$\tan^{-1}(x) \cong 0.999866\, x - 0.3302995\, x^3 + 0.180141\, x^5$$
$$- 0.085133\, x^7 + 0.0208351\, x^9. \qquad (8.32)$$

The identity $\tan^{-1}(x) = -\tan^{-1}(1/x)$ can be employed to compute the arctangent of x if $x \leq 1$.

R8.7   The square root of a positive number x in the range $0.5 \leq x \leq 1$ can be evaluated using the truncated polynomial approximation [Mar92]

$$\sqrt{x} \cong 0.2075806 + 1.454895\, x - 1.34491\, x^2$$
$$+ 1.106812\, x^3 - 0.536499\, x^4 + 0.1121216\, x^5. \qquad (8.33)$$

If x is outside the range from 0.5 to 1, it can be multiplied by a binary constant K^2 to bring the product $x' = K^2 x$ into the desirable range, compute $\sqrt{x'}$ using Eq. (8.25), and then determine $\sqrt{x} = \sqrt{x'}/K$.

8.3   MATLAB Commands Used

The MATLAB commands you will encounter in this exercise are as follows:

General Purpose Commands

        disp   length

Operators and Special Characters

        :   .   +   -   *   /   ;
        %   =   <   ~=

Language Constructs and Debugging

        end   for   function   if

Elementary Matrices and Matrix Manipulation

        fliplr   nargin   pi   :   zeros

Elementary Functions

        abs   cos   log10   real   sin   sqrt

Data Analysis

        conv   max

Two-Dimensional Graphics

```
axis      clf       grid      legend      plot
xlabel    ylabel    stem      title
```

Signal Processing Toolbox

```
butter    buttord    ellip    ellipord    filter
freqz     remez      tf2zp    zp2sos
```

For additional information on these commands, see the *MATLAB Reference Guide* [Mat94] and the *Signal Processing Toolbox User's Guide* [Mat96] or type help commandname in the Command window. A brief explanation of the MATLAB functions used here can be found in Appendix B.

8.4   Simulation of IIR Filters

A causal IIR transfer function of order N is characterized by a transfer function $H(z)$:

$$H(z) = \frac{\sum_{k=0}^{N} p_k\, z^{-k}}{1 + \sum_{k=1}^{N} d_k\, z^{-k}}. \tag{8.34}$$

A number of methods are available for the realization of $H(z)$ resulting in a variety of different structures. Here only some of these structures will be used to demonstrate the simulation of IIR filters on MATLAB. The function filter in the Signal Processing Toolbox of MATLAB simulates a causal IIR filter implemented in the transposed Direct Form II structure as indicated in Figure 6.6(b) for a third-order filter.[3] The basic forms of this function are as follows:

```
y = filter(num,den,x)
[y,sf] = filter(num,den,x,si)
```

The numerator and the denominator coefficients are contained in the vectors num and den, respectively. These vectors do not have to be the same size. The input vector is x while the output vector generated by the filtering algorithm is y. If the first coefficient of den is not equal to 1, the program automatically normalizes all filter coefficients in num and den to make it equal to 1.

In the second form of the function filter, the initial conditions of the delay (state) variables can be specified through the argument si. Moreover, the function filter can return the final values of the delay (state) variables through the output vector sf. The size of the initial (final) condition vector si (sf) is one less than the maximum of the sizes of the filter coefficient vectors num and den. The final values of the state variables given as vector sf are useful if the input vector to be processed is very long and needs to be segmented into

---

[3]See R6.9.

small blocks of data for processing in stages. In such a situation, after the ith block of input data has been processed, the final state vector sf is fed as the initial state vector si for the processing of the $(i + 1)$th block of input data, and so on.

To implement a causal IIR filter implemented in the Direct Form II structure, the function direct2 given below can be employed.

```
function [y,sf] = direct2(p,d,x,si);
% Y = DIRECT2(P,D,X) filters input data vector X with
% the filter described by vectors P and D to create the
% filtered data Y. The filter is a "Direct Form II"
% implementation of the difference equation:
% y(n) = p(1)*x(n) + p(2)*x(n-1) + ...  + p(np+1)*x(n-np)
%        - d(2)*y(n-1) - ...  - d(nd+1)*y(n-nd)
% [Y,SF] = DIRECT2(P,D,X,SI) gives access to initial and
% final conditions, SI and SF, of the delays.
dlen = length(d); plen = length(p);
N = max(dlen,plen); M = length(x);
sf = zeros(1,N-1); y = zeros(1,M);
if nargin ~= 3,
   sf = si;
end
if dlen < plen,
    d = [d zeros(1,plen - dlen)];
    else
    p = [p zeros(1, dlen - plen)];
end
p = p/d(1); d = d/d(1);
for n = 1:M;
    wnew = [1 -d(2:N)]*[x(n) sf]';
    K = [wnew sf];
    y(n) = K*p';
    sf = [wnew sf(1:N-2)];
end
```

For developing the cascade realization of an IIR transfer function, Program P6_1 can be used. Likewise, to develop the parallel realization Program P6_2 can be utilized. In the simulation of a cascade or a parallel-form structure, the individual first- and second-order sections can be realized either in Direct Form II (using function direct2) or in the transposed Direct Form II (using the function filter).

Project 8.1   Structure Simulation and Verification

The structure being simulated can be verified in MATLAB by computing its transfer function with the aid of the function strucver given below.

```
function [p,d] = strucver(ir,N);
H = zeros(2*N+1,N+1);
H(:,1) = ir';
for n = 2:N+1;
    H(:,n) = [zeros(1,n-1) ir(1:2*(N+1)-n)]';
end
H1 = zeros(N+1,N+1);
for k = 1:N+1;
    H1(k,:)  = H(k,:);
end
H3 = zeros(N,N+1);
for k = 1:N;
    H3(k,:)  = H(k+N+1,:);
end
H2 = H3(:,2:N+1);
hf = H3(:,1);
% Compute the denominator coefficients
d = -(inv(H2))*hf;
% Compute the numerator coefficients
p = H1*[1;d];
d = [1; d];
```

Program P8_1 illustrates the design of a causal IIR filter and its simulation in Direct Form II. It employs the functions `strucver` and `direct2` described earlier.

```
% Program P8_1
Wp = [0.4 0.5]; Ws = [0.1 0.8]; Rp = 1; Rs = 30;
[N1, Wn1] = buttord(Wp, Ws, Rp, Rs);
[num,den] = butter(N1,Wn1);
disp('Numerator coefficients are ');disp(num);
disp('Denominator coefficients are ');disp(den);
impres = direct2(num,den,[1 zeros(1,4*N1)]);
[p,d] = strucver(impres,2*N1);
disp('Actual numerator coeffs are '); disp(p');
disp('Actual denominator coeffs are '); disp(d');
```

Questions:

Q8.1 What type of filter is being designed by Program P8_1? What are its specifications? What is the order of the filter? How many impulse response samples are being computed to verify the simulation? Is the simulation correct?

Q8.2 Modify Program P8_1 to simulate the filter in transposed Direct Form II and run the modified prgram. Is the simulation correct?

Q8.3 Develop a cascade realization of the transfer function generated in Question Q8.1 and write a program to simulate it with each individual section implemented in Direct Form

II. Verify the simulation.

Q8.4 Repeat Question Q8.3 with the sections in the cascade in reverse order.

Q8.5 Develop a Parallel Form I realization of the transfer function generated in Question Q8.1 and write a program to simulate it with each individual section implemented in direct form II. Verify the simulation.

Q8.6 Develop a Parallel Form II realization of the transfer function generated in Question Q8.1 and write a program to simulate it with each individual section implemented in transposed Direct Form II. Verify the simulation.

It is quite straightforward to write a MATLAB program to simulate any digital filter structure. We consider next the simulation of the Gray-Markel cascaded lattice realization of an IIR transfer function. Figure 8.1 shows the cascaded lattice realization of the causal third-order transfer function

$$H(z) = \frac{0.44\,z^{-1} + 0.36\,z^{-2} + 0.02\,z^{-3}}{1 + 0.4\,z^{-1} + 0.18\,z^{-2} - 0.2\,z^{-3}} \qquad (8.35)$$

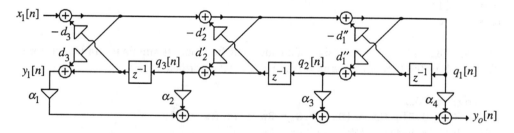

Figure 8.1   Gray-Markel cascaded lattice realization of the IIR transfer function of Eq. (8.35).

The multiplier coefficients of Figure 8.1 are given by:

$$d_3 = -0.2, \quad d_2 = 0.270833, \quad d_1 = 0.35736,$$
$$\alpha_1 = 0.02, \quad \alpha_2 = 0.352, \quad \alpha_3 = 0.276533, \quad \alpha_4 = -0.19016.$$

From Figure 8.1 the equations describing this structure are obtained by inspection as:

$$w_3[n] = x_1[n] - d_3 q_3[n-1],$$
$$w_2[n] = w_3[n] - d_2 q_2[n-1],$$
$$q_1[n] = w_2[n] - d_1 q_1[n-1],$$
$$q_2[n] = d_1 q_1[n] + q_1[n-1],$$
$$q_3[n] = d_2 w_2[n] + q_2[n-1],$$
$$y_1[n] = d_3 w_3[n] + q_3[n-1],$$
$$y_o[n] = \alpha_1 y_1[n] + \alpha_2 q_3[n] + \alpha_3 q_2[n] + \alpha_4 q_1[n].$$

Program P8_2 for the simulation of this structure is given below. The input vector x1 consists of the first seven coefficients of a unit sample sequence. The output vector yo contains the first seven impulse response coefficients, which are then used to determine the numerator and the denominator coefficients of the actual transfer function implemented using the method described in R8.1.

```
% Program P8_2
% Simulation of IIR Cascaded Lattice Structure
%
x1 = [1 zeros(1,6)]; % Generate unit impulse sequence
q3old = 0; q2old = 0; q1old = 0; % Initial conditions
% Enter filter coefficients
D1 = 0.357377; D2 = 0.27083; D3 =-0.2;
alpha1 = 0.02; alpha2 = 0.352;
alpha3 = 0.276533; alpha4 = - 0.19016;
% Compute the first 7 impulse response samples
for n = 1:7
    w3 = x1(n) - D3*q3old;
    w2 = w3 - D2*q2old;
    q1new = w2 - D1*q1old;
    q2new = D1*q1new + q1old;
    q3new = D2*w2 + q2old;
    y1 = D3*w3 + q3old;
    yo(n) = alpha1*y1 + alpha2*q3new + alpha3*q2new + alpha4*q1new;
    q3old = q3new; q2old = q2new; q1old = q1new;
end
[num,den] = strucver(yo,3);
disp('Numerator coefficients are');disp(num');
disp('Denominator coefficients are');disp(den')
```

Questions:

Q8.7 Run Program P8_2 and generate the transfer function of the cascaded lattice structure being simulated. Is the transfer function generated the same as in Eq. (8.35)?

Q8.8 Using MATLAB, generate the transfer function of an elliptic lowpass filter with the following specifications: passband edge at 0.4π, stopband edge at 0.6π, passband ripple of 0.5 dB, and minimum stopband attenuation of 28 dB. Using Program P6_4 develop its Gray-Markel cascaded lattice realization. Simulate and verify the realization using Program P8_2.

Project 8.2 Illustration of Filtering

Program P8_3 illustrates the design of a causal IIR filter, its simulation in transposed Direct Form II, and its application in filtering a signal.

```
% Program P8_3
% Illustration of Filtering by an IIR Filter
%
clf;
% Generate the input sequence
k = 0:50;
w2 = 0.7*pi;w1 = 0.2*pi;
x1 = 1.5*cos(w1*k); x2 = 2*cos(w2*k);
x = x1+x2;
% Determine the filter transfer function
[N, Wn] = ellipord(0.25, 0.55, 0.5, 50);
[num, den] = ellip(N,0.5, 50,Wn);
% Generate the output sequence
y = filter(num,den,x);
% Plot the input and the output sequences
subplot(2,1,1);
stem(k,x); axis([0 50 -4 4]);
xlabel('Time index n'); ylabel('Amplitude');
title('Input Sequence');
subplot(2,1,2);
stem(k,y); axis([0 50 -4 4]);
xlabel('Time index n'); ylabel('Amplitude');
title('Output Sequence');
```

Questions:

Q8.9 What type of filter is being designed by Program P8_3? What are its specifications? What is the order of the filter? What are the frequencies of the sinusoidal sequences forming the input?

Q8.10 Run Program P8_3 and generate the two plots. Which component of the input appears at the filter output? Why is the beginning part of the output sequence not a perfect sinusoid? Modify Program P8_3 to filter the sequence $x2[n]$ only. Is the output generated as expected? Justify your answers.

The cascade form of an IIR transfer function can be generated from its zero-pole description using the function zp2sos. The code fragment given below illustrates the generation of the transfer functions of each section of the elliptic lowpass transfer function example of Program P8_3 and its implementation in cascade form along with the overall transfer function as one section with all realizations in Direct Form II.

```
[N, Wn] = ellipord(0.25, 0.55, 0.5, 50);
[num,den] = ellip(N,0.5, 50,Wn);
[z,p,const] = tf2zp(num,den);
sos = zp2sos(z,p,const);
row1 = real(sos(1,:));row2 = real(sos(2,:));
num1 = row1(1:3);den1 = row1(4:6);
```

```
num2 = row2(1:3);den2 = row2(4:6);
y = direct2(num,den,x);
y1 = direct2(num1,den1,x);y2 = direct2(num2,den2,y1);
```

Questions:

Q8.11   Using the above code fragment modify Program P8_3 to filter the sequence being generated by a cascade structure and compare the output generated with the output generated when filtered by a single higher-order section. Is there any difference between the two outputs? Show precisely the two filters being simulated by writing down the expression for the transfer function of each simulation.

Q8.12   Using the function `strucver` modify the program developed in Question Q8.11 to verify the structure being simulated. Run the modified program. Are your simulations correct?

A long input sequence can be filtered using the overlap-add method in which the input sequence is segmented into a set of contiguous short input blocks, each block is then filtered separately, and the overlaps in the output blocks are added appropriately to generate the long output sequence. This method of filtering can be implemented easily on MATLAB using the second form of the function `filter`. Here the final values of the state variable vector `sf` at any stage of filtering are fed back in the following stage of filtering as the initial condition vector `si`.

Question:

Q8.13   Modify Program P8_3 by filtering the input sequence into a set of contiguous blocks of length 5 each. Run the modified program and compare your result with that generated by filtering the input as one segment.

8.5   Simulation of FIR Digital Filters

The functions `direct2` and `filter` can also be used to implement an FIR digital filter as illustrated in this project.

Project 8.3   Structure Simulation and Verification

Program P8_4 illustrates the design of a causal FIR filter and its simulation in transposed Direct Form II.

```
% Program P8_4
num = remez(9, [0 0.3 0.5 1],[1 1 0 0]);
disp('Filter coefficients are ');disp(num);
impres = filter(num,1 ,[1 zeros(1,9)]);
disp('Actual filter coeffs are '); disp(impres);
```

Questions:

Q8.14 What type of filter is being designed by Program P8_4? What are its specifications? What is the order of the filter? How many impulse response samples are being computed to verify the simulation? Is the simulation correct?

Q8.15 Modify Program P8_4 to simulate the filter in Direct Form II and run the modified program. Is the simulation correct?

Q8.16 Develop a cascade realization of the transfer function generated in Question Q8.14 by Program P8_4 and write a program to simulate it with each individual section implemented in Direct Form II. Verify the simulation.

Q8.17 Repeat Question Q8.16 with the sections in the cascade in reverse order.

Project 8.4   Illustration of Filtering

Question:

Q8.18 Modify Program P8_3 to simulate the direct-form realization of the FIR filter of Program P8_4 using the function direct2 and demonstrate its filtering properties.

8.6   Design of Tunable Digital Filters

In the next two projects you will investigate the design of tunable IIR and FIR digital filters.

Project 8.5   Design of Tunable IIR Filter

Program P_5 illustrates the design of a tunable causal IIR lowpass filter based on the parallel allpass realization (see R6.19).

```
% Program P8_5
% Illustration of Tunable IIR Filter Design
clf;
[z,p,k] = ellip(5,0.5,40,0.4);
a = conv([1 -p(1)],[1 -p(2)]);b = [1 -p(5)];
c = conv([1 -p(3)],[1 -p(4)]);
w = 0:pi/255:pi;
alpha = [0 0.1 -0.25];
for i = 1:3
    an1 = a(2) + (a(2)*a(2) - 2*(1 + a(3)))*alpha(i);
    an2 = a(3) + (a(3) -1)*a(2)*alpha(i);
    g = b(2) - (1 - b(2)*b(2))*alpha(i);
    cn1 = c(2) + (c(2)*c(2) - 2*(1 + c(3)))*alpha(i);
    cn2 = c(3) + (c(3) -1)*c(2)*alpha(i);
    a = [1 an1 an2];b = [1 g]; c = [1 cn1 cn2];
    h1 = freqz(fliplr(a),a,w); h2 = freqz(fliplr(b),b,w);
```

```
    h3 = freqz(fliplr(c),c,w);
    ma(i,:)  = 20*log10(abs(0.5*(h1.*h2 + h3)));
end
plot(w/pi,ma(1,:),'r-',w/pi,ma(2,:),'b--',w/pi,ma(3,:),'g-.');
grid;
axis([0 1 -80 5]);
xlabel('\omega/\pi');ylabel('Gain, dB');
legend('\alpha = 0 ','\alpha = 0.1','\alpha = -0.25');
```

Questions:

Q8.19 What type of filter is being designed as the nominal filter by Program P8_5? What are its specifications? What is the order of the filter?

Q8.20 Write a MATLAB program to design this nominal filter and display its pole locations using the command zplane. Determine the transfer functions of the two allpass sections.

Q8.21 Run Program P8_5 and display all gain responses. What are the cutoff frequencies of the filters being designed?

Q8.22 Modify Program P8_5 to determine and plot the gain response of the power-complementary filter of each of the filters being designed in Program P8_5.

Project 8.6   Design of Tunable FIR Filter

Program P8_6 illustrates the design of a tunable causal FIR lowpass filter based on the method outlined in Section R8.3.

```
% Program 8_6
% Illustration of Tunable FIR Filter Design
clf;
w = 0:pi/255:pi;
f = [0 0.36 0.46 1];m = [1 1 0 0];
b1 = remez(50, f, m);
h1 = freqz(b1,1,w);
m1 = 20*log10(abs(h1));
n = -25:-1;
c = b1(1:25)./sin(0.41*pi*n);
wc = [0.31*pi 0.51*pi];
for i = 1:2
    d = c.*sin(wc(i)*n);
    q = (b1(26)*wc(i))/(0.4*pi);
    b2 = [d q fliplr(d)];
    mag(i,:)  = 20*log10(abs(freqz(b2,1,w)));
end
plot(w/pi,mag(1,:),'b--',w/pi,m1,'r-',w/pi,mag(2,:),'g-.');
```

```
grid;
axis([0 1 -80 5]);
xlabel('\omega/\pi');ylabel('Gain, dB');
legend('\omega_{c} = 0.31 \pi','\omega_{c} = 0.41\pi','\omega_{c} =
0.51\pi')
```

Questions:

Q8.23   What type of filter is being designed as the nominal filter by Program P8_6? What are its specifications? What is the order of the filter?

Q8.24   Run Program P8_6 and display all gain responses. What are the cutoff frequencies of the filters being designed?

Q8.25   Modify Program P8_6 to determine and plot the gain response of the delay-complementary filter of each of the filters being designed in Program P8_6.

8.7   DFT Computation

Project 8.7   Computational Complexity of Algorithms

Question:

Q8.26   Write a MATLAB program to compute and display the N-point DFT of a length-N sequence using directly Eq. (8.26) and using the M-function `fft`. Your program should also determine the total number of floating point operations involved in the computation of the DFT using the command `flops` for both cases. The input data are the sequence entered in a vector form from which the length N of the sequence is determined. Using this program, compute the DFTs of sequences of your choice of lengths 64, 128, and 256. Comment on your results.

8.8   Function Approximation

Project 8.8   Computation of Trigonometric and Other Functions

Questions:

Q8.27   Write a MATLAB program to compute and plot $\sin(x)$ as a function of x using the approximation of Eq. (8.31) where x is in radians and in the range $0 \leq x \leq \pi/2$. Plot also the error due to the approximation. Run this program and generate the two plots. Comment on your results.

Q8.28   Write a MATLAB program to compute and plot $\tan^{-1}(x)$ as a function of x using the approximation of Eq. (8.32) where x is in the range $0 \leq x \leq 1$. Plot also the error due to the approximation. Run this program and generate the two plots. Comment on your results.

Q8.29 Write a MATLAB program to plot the error due to the computation of $\sqrt{x}$ using the approximation of Eq. (8.33) as a function of x where x is in the range $0.5 \leq x \leq 1$. Run this program and generate the two plots. Comment on your results.

8.9 Background Reading

[1] S.K. Mitra. *Digital Signal Processing: A Computer-Based Approach.* McGraw-Hill, New York NY, 1998. Secs. 8.1, 8.2, 8.7, 8.8.

Analysis of Finite Word-Length Effects

9.1 Introduction

When implemented in either software form on a general-purpose computer or in special-purpose hardware form, the parameters of the LTI discrete-time system along with the signal variables can take only discrete values within a specified range since the registers of the digital machine where they are stored are of finite length. If the quantization amounts are small compared to the values of the signal variables and filter constants, a simpler approximate theory based on a statistical model can be applied and it is possible to derive the effects of discretization and develop results that can be verified experimentally. One of the sources of quantization errors is caused by the quantization of the multiplier coefficients characterizing the filter structure realizing the given transfer function. In the case of digital processing of continuous-time signals, a second error source is caused by the A/D conversion process. The quantization of arithmetic operations leads to a third source of errors. Another type of error occurs in digital filters due to the nonlinearity caused by the quantization of arithmetic operations. This last source of errors may result in oscillations at the filter's output, called *limit cycles*, usually in the absence of input or sometimes in the presence of constant input signals or sinusoidal input signals. An analysis of the various quantization effects on the performance of a digital filter in practice depends on whether the numbers are in fixed-point or floating-point format, the type of representation for the negative numbers being used, the quantization method being employed to quantize the data, and the digital filter structure being used for implementation. In this exercise you will study the effects of the above sources of quantization errors in digital filter structures implemented using fixed-point arthmetic and investigate structures that are less sensitive to these effects.

9.2 Background Review

R9.1 To accommodate the representation of both positive and negative b-bit fractions, an additional bit, called the *sign bit*, is placed at the leading position of the register to indicate the sign of the number. Independent of the scheme being used to represent the negative number, the sign bit is 0 for a positive number and 1 for a negative number.

R9.2 Two types of quantization are employed to represent a fixed-point number in a register of finite word-length: truncation and rounding. *Truncation* of a fixed-point number from $(\beta + 1)$ bits to $(b + 1)$ bits is implemented by simply discarding the least significant $(\beta - b)$ bits. In the case of *rounding*, the number is quantized to the nearest quantization level, and a number exactly halfway between two quantization levels is assumed to be rounded up to the nearest higher level. Therefore, if the bit $a_{-(b+1)}$ is 0, rounding is equivalent to truncation and if this bit is 1, 1 is added to the least-significant-bit position of the truncated number.

R9.3 The transfer function $\hat{H}(z)$ of the digital filter implemented with quantized multiplier coefficients either in hardware or software form is different from the desired transfer function $H(z)$. The main effect of the coefficient quantization is therefore on the poles and zeros, which move to different locations than the original desired locations. As a result, the actual frequency response $\hat{H}(e^{j\omega})$ is different from the desired frequency response $H(e^{j\omega})$. Moreover, the poles may move outside the unit circle, causing the implemented digital filter to become unstable even though the original transfer function with unquantized coefficients is stable.

R9.4 As the input-output characteristic of an A/D converter is nonlinear and the analog input signal, in most practical cases, is not known a priori, for analy sis purposes it is assumed that the quantization error $e[n]$ is a random signal and a statistical model of the quantizer operation as indicated in Figure 9.1 is employed.

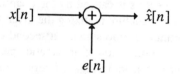

Figure 9.1 A statistical model of the A/D quantizer.

In addition, the following assumptions are made for a simplified analysis:

1. The error sequence $\{e[n]\}$ is a sample sequence of a wide-sense stationary (WSS) white noise process with each sample $e[n]$ being uniformly distributed over the range of the quantization error as indicated in Figure 9.2 where δ is the quantization step.
2. The error sequence is uncorrelated with its corresponding input sequence $\{x[n]\}$.
3. The input sequence is a sample sequence of a stationary random process.

For a fixed-point fraction of length $(b+1)$ bits (with one bit assigned for the sign), $\delta = 2^{-b}$.

In the case of ones'-complement or sign-magnitude truncation, the quantization error is correlated to the input signal, as here the sign of each error sample $e[n]$ is exactly opposite to the sign of the corresponding input sample $x[n]$. As a result, in digital signal processing either rounding or two's-complement truncation is employed for quantizing a number. The mean and the variance of the error sample in the case of rounding are given by

$$m_e = \frac{(\delta/2) - (\delta/2)}{2} = 0, \tag{9.1}$$

$$\sigma_e^2 = \frac{[(\delta/2) - (-\delta/2)]^2}{12} = \frac{\delta^2}{12}. \tag{9.2}$$

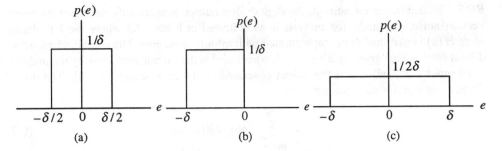

Figure 9.2 Quantization error probability density functions: (a) rounding, (b) two's-complement truncation, and (c) ones'-complement truncation.

The corresponding parameters for the two's-complement truncation are as follows:

$$m_e = \frac{0 - \delta}{2} = -\frac{\delta}{2}, \tag{9.3}$$

$$\sigma_e^2 = \frac{(0 - \delta)^2}{12} = \frac{\delta^2}{12}. \tag{9.4}$$

The effect of the additive quantization noise $e[n]$ on the input signal $x[n]$ of an A/D converter is given by the *signal-to-quantization noise ratio* $(SNR_{A/D})$, defined by

$$SNR_{A/D} = 10 \log_{10} \left(\frac{\sigma_x^2}{\sigma_e^2} \right) \text{ dB}, \tag{9.5}$$

where σ_x^2 is the input signal variance representing the signal power and σ_e^2 is the noise variance representing the quantization noise power. For rounding, the quantization error is uniformly distributed in the range $(-\delta/2, \delta/2)$ and for two's-complement truncation, the quantization error is uniformly distributed in the range $(-\delta, 0)$ as indicated in Figures 9.2(a) and (b), respectively. In the case of a bipolar $(b + 1)$-bit A/D converter $\delta = 2^{-(b+1)} R_{FS}$, where R_{FS} is the full-scale range of the converter and, hence,

$$SNR_{A/D} = 10 \log_{10} \left(\frac{48 \, \sigma_x^2}{2^{-2b}(R_{FS})^2} \right) = 6.02 \, b + 16.81 - 20 \log_{10} \left(\frac{R_{FS}}{\sigma_x} \right) \text{ dB.} \tag{9.6}$$

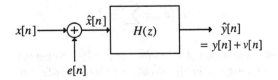

Figure 9.3 Model for the analysis of the effect of processing a quantized input by an LTI discrete-time system.

R9.5   To determine the noise at the digital filter output generated by the input A/D conversion noise, the model for analysis is as indicated in Figure 9.3 where the LTI digital filter $H(z)$ is assumed to be implemented with infinite precision. The actual output of the digital filter is thus given by $y[n] + v[n]$, where $y[n]$ is the output generated by the unquantized input $x[n]$, and $v[n]$ is the output generated by the error sequence $e[n]$. The output noise component $v[n]$ is thus given by

$$v[n] = \sum_{m=-\infty}^{\infty} e[m] \, h[n-m], \tag{9.7}$$

where $h[n]$ is the impulse response of the digital filter. The mean m_v and the variance σ_v^2 of the output noise $v[n]$ are then given by

$$m_v = m_e \, H(e^{j0}), \tag{9.8}$$

$$\sigma_v^2 = \frac{\sigma_e^2}{2\pi} \int_{-\pi}^{\pi} |H(e^{j\omega})|^2 \, d\omega. \tag{9.9}$$

The output noise power spectrum is given by

$$P_{vv}(\omega) = \sigma_e^2 \, |H(e^{j\omega})|^2, \tag{9.10}$$

and the normalized output noise variance is given by

$$\sigma_{v,n}^2 = \frac{\sigma_v^2}{\sigma_e^2} = \frac{1}{2\pi} \int_{-\pi}^{\pi} |H(e^{j\omega})|^2 \, d\omega \tag{9.11}$$

$$= \frac{1}{2\pi j} \oint_C H(z) \, H(z^{-1}) \, z^{-1} dz, \tag{9.12}$$

where C is a counterclockwise contour in the ROC of $H(z) \, H(z^{-1})$. An equivalent expression for Eq. (9.11) is given by

$$\sigma_{v,n}^2 = \sum_{n=-\infty}^{\infty} |h[n]|^2. \tag{9.13}$$

R9.6   A simple algebraic method for computing the normalized output noise variance of a causal stable real rational function $H(z)$ with simple poles is based on a partial-fraction form expansion of $H(z)$ [Mit74b]:

$$H(z) = \sum_{k=1}^{L} H_k(z), \tag{9.14}$$

where $H_k(z)$ is either a constant A or of the form $\frac{B_k}{z - z_k}$. Substituting Eq. (9.14) into Eq. (9.12), the expression for the normalized output noise variance can be rewritten as

$$\sigma_{v,n}^2 = \frac{1}{2\pi j} \left\{ \sum_{k=1}^{R} \oint_C H_k(z) \, H_k(z^{-1}) \, z^{-1} \, dz + 2 \sum_{k=1}^{R-1} \sum_{\ell=k+1}^{R} \oint_C H_k(z) \, H_\ell(z^{-1}) \, z^{-1} \, dz \right\}. \tag{9.15}$$

Denote a typical contour integral in Eq. (9.15) as

$$I_i = \frac{1}{2\pi j} \oint_C H_k(z) \, H_\ell(z^{-1}) \, z^{-1} \, dz. \qquad (9.16)$$

The expressions for different I_i are listed in Table 9.1.

Table 9.1 Expressions for typical contour integrals.

$H_k(z)$	$H_\ell(z^{-1})$	
	A	$\dfrac{B_\ell}{z^{-1}-a_\ell}$
A	$I_1 = A^2$	0
$\dfrac{B_k}{z-a_k}$	0	$I_2 = \dfrac{B_k B_\ell}{1-a_k a_\ell}$

R9.7 For a causal and stable digital filter, the impulse response decays rapidly to zero values, and hence Eq. (9.13) can be approximated as a finite sum

$$\sigma_{v,n}^2 = \sum_{n=0}^{L} |h[n]|^2. \qquad (9.17)$$

A convenient and practical approach to determine an appropriate value of L is based on an iterative computation of the above partial sum for $L = 1, 2, \ldots$, and then stopping the computation when the $|h[L]|^2$ becomes smaller than a specified value κ, which is typically chosen as 10^{-7}.

R9.8 In the fixed-point implementation of a digital filter, only the result of a multiplication operation is quantized. The statistical model for the analysis of product roundoff errors is indicated in Figure 9.4. Here the output $v[n]$ of the ideal multiplier is quantized to a value $\hat{v}[n]$, where $\hat{v}[n] = v[n] + e_\alpha[n]$.

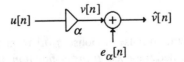

Figure 9.4 Statistical model for the product round-off error analysis.

For analysis purposes the following assumptions are made:

1. The error sequence $\{e_\alpha[n]\}$ is a sample sequence of a stationary white noise process with each sample $e_\alpha[n]$ being uniformly distributed over the range of the quantization error.

2. The quantization error sequence $e_\alpha[n]$ is uncorrelated with the sequence $\{v[n]\}$, the input sequence $\{x[n]\}$ to the digital filter, and all other quantization noise sources.

The assumption of $\{e_\alpha[n]\}$ being uncorrelated with $\{v[n]\}$ holds only for rounding and two's-complement truncation. The mean and variance of the error sample for rounding are given by Eqs. (9.1) and (9.2), respectively, while those for the two's-complement truncation are given by Eqs. (9.3) and (9.4), respectively.

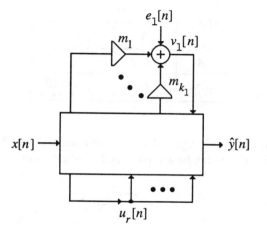

Figure 9.5   The statistical model of a digital filter structure for product round-off error analysis.

R9.9   The statistical model of a digital filter for the product round-off error analysis is as indicated in Figure 9.5, which explicitly shows the quantized outputs of the k_ℓ multipliers at its input. This figure also shows the internal rth branch node associated with the signal variable $u_r[n]$, which needs to be scaled to prevent overflows at these nodes. The error sources are assumed to be statistically independent of each other and, as a result, the total round-off noise at the output of the digital filter is the sum of the noise generated by each noise source.

The z-transform $G_\ell(z)$ of the impulse response $g_\ell[n]$ from the input of the ℓth adder to the digital filter output is called the *noise transfer function*, and the z-transform $F_r(z)$ of the impulse response $f_r[n]$ of the impulse response from the digital filter input to the rth branch node is called the *scaling transfer function*.

The output noise variance caused by the noise source $e_\ell[n]$ is given by

$$\sigma_o^2 \left[k_\ell \left(\frac{1}{2\pi j} \oint_C G_\ell(z)\, G_\ell(z^{-1})\, z^{-1}\, dz \right) \right] = \sigma_o^2 \left[k_\ell \left(\frac{1}{2\pi} \int_{-\pi}^{\pi} |G_\ell(e^{j\omega})|^2\, d\omega \right) \right],$$

$$(9.18)$$

where σ_o^2 denotes the variance of the individual noise source at the output of each multiplier. The total output noise power due to all product round-offs is given by

$$\sigma_\gamma^2 = \sigma_o^2 \sum_{\ell=1}^{L} k_\ell \left(\frac{1}{2\pi j} \oint_C G_\ell(z)\, G_\ell(z^{-1})\, z^{-1}\, dz \right), \tag{9.19}$$

where L is the total number of such adders in the filter structure. In structures where the quantization operation is carried out after all of the multiply-add operations have been completed, $k_\ell = 1$.

R9.10 In a digital filter implemented using fixed-point arithmetic, overflow may occur at certain internal nodes, which may lead to large amplitude oscillations at the filter output causing unsatisfactory operation. The probability of overflow is minimized significantly by properly scaling the internal signal levels with the aid of scaling multipliers inserted at appropriate points in the digital filter structure.

With reference to Figure 9.5, the objective of scaling is to ensure that

$$|u_r[n]| \leq 1, \quad \text{for all } r \text{ and for all values of } n, \tag{9.20}$$

assuming all fixed-point numbers are represented as binary fractions and the input sequence of the filter is bounded by unity, that is,

$$|x[n]| \leq 1, \quad \text{for all values of } n. \tag{9.21}$$

A general scaling rule is given by [Jac70]

$$|u_r[n]| \leq \|F_r\|_q \cdot \|X\|_p, \tag{9.22}$$

or all $p, q \geq 1$ satisfying $\frac{1}{p} + \frac{1}{q} = 1$. In Eq. (9.22), $\|X\|_p$ denotes the $\mathcal{L}_p$-norm ($p \geq 1$) of a Fourier transform $X(e^{j\omega})$ defined by

$$\|X\|_p \triangleq \left(\frac{1}{2\pi} \int_{-\pi}^{\pi} |X(e^{j\omega})|^p d\omega \right)^{\frac{1}{p}}. \tag{9.23}$$

Note that for the $\mathcal{L}_\infty$-bound, $p = \infty$ and $q = 1$, and for the $\mathcal{L}_2$-bound, $p = q = 2$. Another useful scaling rule, $\mathcal{L}_1$-bound, is obtained for $p = 1$ and $q = \infty$.

After scaling, the scaling transfer functions become $\|\bar{F}_r\|_q$ and the scaling constants are chosen such that

$$\|\bar{F}_r\|_q \leq 1, \quad r = 1, 2, \ldots, R. \tag{9.24}$$

R9.11 A number of digital filter realization techniques have been proposed that result in structures that are inherently less sensitive to quantization of the multiplier coefficients. A key requirement for low sensitivity realization is that the prescribed transfer function $H(z)$ be a *bounded real* (BR) function; that is, $H(z)$ is a causal stable real coefficient function characterized by a magnitude response $|H(e^{j\omega})|$ bounded above by unity, that is,

$$|H(e^{j\omega})| \leq 1. \tag{9.25}$$

It is also assumed that at a set of frequencies, ω_k, the magnitude of $H(z)$ is exactly equal to 1, that is,

$$|H(e^{j\omega_k})| = 1. \tag{9.26}$$

The frequencies ω_k are in the passband of the filter as the magnitude response is bounded above by unity. The digital filter structure realizing a BR transfer function exhibits zero sensitivity at the set of frequencies ω_k in the passband and low sensitivity at other frequencies in the passband if the transfer function of the structure realized with quantized multiplier coefficients (assuming small changes in the coefficient values) remains bounded-real, satisfying the condition of Eq. (9.25). Such a structure is said to be *structurally passive*.

R9.12   The parallel allpass realization of a bounded-real transfer function discussed in Section R6.19 satisfies the low passband sensitivity and is thus structurally passive.

A bounded-real Type 1 FIR transfer function $H(z)$ of degree N can be realized in a structurally passive form by expressing its delay-complementary transfer function $G(z)$ in the form of a cascade of two FIR filter sections $G_a(z)$ and $G_b(z)$, where $G_b(z)$ has all zeros on the unit circle with multiplicity 2:

$$G(z) = G_a(z)\,G_b(z) = G_a(z) \prod_{k=1}^{L}(1 - 2\,\cos\omega_k z^{-1} + z^{-2})^2, \tag{9.27}$$

and then realizing $G_b(z)$ as a cascade of $2L$ second-order sections with multiplier coefficients $2\cos\omega_k$. Finally, the transfer function $H(z)$ is realized in the form

$$H(z) = z^{-N/2} - G_a(z)\,G_b(z), \tag{9.28}$$

as indicated in Figure 9.6.

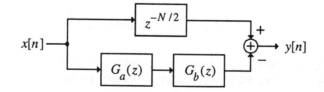

Figure 9.6   Low passband sensitivity realization of a Type 1 FIR filter $H(z)$.

R9.13   A practical digital filter is a nonlinear system caused by the quantization of the arithmetic operations. As a result, an IIR filter, which is stable under infinite precision, may exhibit an unstable behavior under finite precision arithmetic for specific input signals, such as zero or constant inputs. This type of instability usually results in an oscillatory periodic output called a *limit cycle*. There are basically two types of limit cycles: (1) *granular* and (2) *overflow*. The former type of limit cycles is usually of low amplitude, whereas overflow oscillations have large amplitudes.

9.3   MATLAB Commands Used

The MATLAB commands you will encounter in this exercise are as follows:

General Purpose Commands

    `disp`

Operators and Special Characters

    `:`   `.`   `+`   `-`   `*`   `/`   `;`
    `%`   `<`

Language Constructs and Debugging

    `break`   `end`   `for`   `if`   `input`

Elementary Matrices and Matrix Manipulation

    `ones`   `pi`   `zeros`

Elementary Functions

    `abs`   `cos`

Polynomial and Interpolation Functions

    `conv`

Two-Dimensional Graphics

    `axis`   `plot`   `stem`   `title`   `xlabel`
    `ylabel`

General Purpose Graphics Functions

    `clf`   `subplot`

Character String Functions

    `num2str`

Signal Processing Toolbox

```
filter          impz
```

For additional information on these commands, see the *MATLAB Reference Guide* [Mat94] and the *Signal Processing Toolbox User's Guide* [Mat96] or type help commandname in the Command window. A brief explanation of the MATLAB functions used here can be found in Appendix B.

9.4   Generation and Quantization of Binary Numbers

As MATLAB uses decimal numbers and arithmetic to study the quantization effects on digital filters implemented using binary numbers and arithmetic, it is convenient to develop the decimal equivalent of quantized representations of binary numbers and signals.

Project 9.1   Generation of Decimal Equivalent of Quantized Binary Numbers

The aim of this project is to generate the decimal equivalent of the binary representation in sign-magnitude form of a decimal number with a specified number of bits for the fractional part obtained by truncation. To this end, the MATLAB function a2dT given below can be used.

```
function beq = a2dT(d,n)
% BEQ = A2DT(D, N) generates the decimal
% equivalent BEQ of the binary representation
% of a vector D of decimal numbers with N bits
% for the magnitude part obtained by truncation
%
m = 1; d1 = abs(d);
while fix(d1) > 0
    d1 = abs(d)/(10^m);
    m = m+1;
end
beq = 0;
for k = 1:n
    beq = fix(d1*2)/(2^k) + beq;
    d1 = (d1*2) - fix(d1*2);
end
beq = sign(d).*beq*10^(m-1);
```

Questions:

Q9.1   Explain the operation of the function a2dT. What is the purpose of the command fix? Write a MATLAB program to convert an arbitrary decimal number into its quantized equivalent employing the function a2dT and display the equivalent representation. The input data are the decimal number d to be converted and the number of bits N assigned for the fractional part, and the output is the decimal equivalent beq of the quantized binary number.

Q9.2 Using the above program generate the binary equivalents in sign-magnitude form of the following decimal numbers: (1) 5.3749, (2) − 21.78239, (3) 0.79889. Use six bits for the fractional part.

Q9.3 Repeat Question Q9.2 with eight bits for the fractional part.

Q9.4 Develop the ones'-complement and two's-complement representations of the binary numbers generated in Questions Q9.2 and Q9.3.

The MATLAB function a2dR given below can be used to generate the decimal equivalent of the binary representation in sign-magnitude form of a decimal number with a specified number of bits for the fractional part obtained by rounding.

```
function beq = a2dR(d,n)
% BEQ = A2DR(D, N) generates the decimal
% equivalent BEQ of the binary representation
% of a vector D of decimal numbers with N bits
% for the magnitude part obtained by rounding
%
m = 1; d1 = abs(d);
while fix(d1) > 0
    d1 = abs(d)/(10^m);
    m = m+1;
end
beq = 0;d1 = d1 + 2^(-n-1);
for k = 1:n
    beq = fix(d1*2)/(2^k) + beq;
    d1 = (d1*2) - fix(d1*2);
end
beq = sign(d).*beq*10^(m-1);
```

Question:

Q9.5 What is the difference between the functions a2dR and a2dT? How is the rounding being performed?

9.5 Coefficient Quantization Effects

As indicated earlier (see R9.4), the transfer function of the digital filter realized with quantized multiplier coefficients is in general different from the desired transfer function that causes the poles and zeros to move from their desired locations, resulting in a frequency response that is different from the desired one. In this section you will study these effects.

Project 9.2   Effect on Frequency Response and Pole-Zero Locations

You will first investigate the effect of multiplier coefficient quantizations on the direct form realization of an IIR transfer function. To this end, Program P9_1 given below can be employed. The program uses the M-function pzplot, which is same as the M-function zplane except here the poles are shown with the symbol + and the zeros are shown with the symbol * in the pole-zero plot.[1]

```
% Program P9_1
% Coefficient Quantization Effects on Direct Form
% Realization of an IIR Transfer Function
clf;
[b,a] = ellip(6,0.05,60,0.4);
[g,w] = gain(b,a);
bq = a2dT(b,6);aq = a2dT(a,6);
[gq,w] = gain(bq,aq);
plot(w/pi,g,'b', w/pi,gq,'r--');
axis([0 1 -80 1]);grid
xlabel('\omega/\pi');ylabel('Gain, dB');
title('original - solid line; quantized - dashed line');
pause
zplane(b,a);
hold on;
pzplot(bq,aq);
title('Original pole-zero locations:  x, o; New pole-zero locations:
+, *')
```

Questions:

Q9.6   Which statement determines the IIR transfer function? What are the order of the transfer function and its type? Which statements determine the decimal equivalents of the quantized binary representations of the transfer function coefficients? How many bits are being assigned to the fractional part of the binary representations?

Q9.7   Run the above program and generate the two plots. Comment on your results.

Q9.8   Modify Program P9_1 to investigate the coefficient effects on an eighth order elliptic highpass transfer function with a passband ripple of 0.1 dB, a minimum stopband attenuation of 70 dB, and a normalized cutoff frequency at 0.55 rad/sec. Assign five bits to the fractional part of the binary representations. Run the modified program. Comment on your results.

It is of interest to compare the performance of the direct form realization of an IIR transfer function with that of a cascade realization with both realized with quantized coefficients. Program P9_2 evaluates the effect of multiplier coefficient quantizations on the cascade form realization.

---

[1]The modification to the function zplane is with permission from The Mathworks, Inc., Natick, MA.

```
% Program P9_2
% Coefficient Quantization Effects on Cascade
% Realization of an IIR Transfer Function
clf;
[z,p,k] = ellip(6,0.05,60,0.4);
[b,a] = zp2tf(z,p,k);
[g,w] = gain(b,a);
sos = zp2sos(z,p,k);
sosq = a2dT(sos,6);
R1 = sosq(1,:);R2 = sosq(2,:);R3 = sosq(3,:);
b1 = conv(R1(1:3),R2(1:3));bq = conv(R3(1:3),b1);
a1 = conv(R1(4:6),R2(4:6));aq = conv(R3(4:6),a1);
[gq,w] = gain(bq,aq);
plot(w/pi,g,'b', w/pi,gq,'r--');
axis([0 1 -80 1]);grid
xlabel('\omega/\pi');ylabel('Gain, dB');
title('original - solid line; quantized - dashed line');
pause
zplane(b,a);
hold on;
pzplot(bq,aq);
title('Original pole-zero locations:  x, o; New pole-zero locations:
+, *')
```

Questions:

Q9.9   What are the order of the transfer function generated by Program P9_2 and its type? What is the function of the command zp2sos? Which statement determines the decimal equivalents of the quantized binary representations of the transfer function coefficients? How many bits are being assigned to the fractional part of the binary representations?

Q9.10   Run the above program and generate the two plots. Compare the plots generated by Program P9_2 with those generated in Question Q9.6. What can you conclude from this comparison?

Q9.11   Modify Program P9_2 to investigate the coefficient effects on an eighth order elliptic highpass transfer function with a passband ripple of 0.1 dB, a minimum stopband attenuation of 70 dB, and a normalized cutoff frequency at 0.55 rad/sec. Assign five bits to the fractional part of the binary representations. Run the modified program and comment on your results.

The above two programs can be modified easily to investigate the multiplier coefficient effects on an FIR transfer function, as illustrated by Program P9_3 given below for the direct form realization.

```
% Program P9_3
```

```
% Coefficient Quantization Effects on Direct-Form
% Realization of an FIR Transfer Function
clf;
f = [0 0.4 0.45 1]; m = [1 1 0 0];
b = remez(19, f, m);
[g,w] = gain(b,1);
bq = a2dT(b,5);
[gq,w] = gain(bq,1);
plot(w/pi,g,'b-', w/pi,gq,'r--');
axis([0 1 -60 5]);grid
xlabel('\omega/\pi');ylabel('Gain, dB');
title('original - solid line; quantized - dashed line');
```

Questions:

Q9.12   What is the order of the transfer function generated by Program P9_3 and its type? What are its desired magnitude response specifications?

Q9.13   Run the above program and generate the magnitude response plots. Comment on your results.

Q9.14   Modify Program P9_3 to investigate the coefficient effects on a 25th order equiripple highpass transfer function with stopband edges at 0 and 0.6, and passsband edges at 0.65 and 1. Assign four bits to the fractional part of the binary representations. Run the modified program and comment on your results.

9.6   A/D Conversion Noise Analysis

Project 9.3   Evaluation of A/D Signal-to-Quantization Noise Ratio

In this project you will investigate first the relation between the A/D converter wordlength b and the signal-to-quantization noise ratio $SNR_{A/D}$ using MATLAB. The basis for the SNR computation is Eq. (9.6).

Question:

Q9.15   Write a MATLAB program to determine the signal-to-quantization noise ratio in the digital equivalent of an analog sample $x[n]$ with a zero-mean Gaussian distribution using a $(b + 1)$-bit A/D converter (with one bit assigned for the sign) having a full-scale range $R_{FS} = K\sigma_x$. Using this program compute for the following values of b: 7, 9, 11, 13, and 15, and for the following values of K: 4, 6, and 8.

Project 9.4   Computation of Output Noise Variance

The aim of this project is to investigate the propagation of input quantization noise to the output of a causal, stable LTI digital filter. To this end, the function noisepwr1 given below and based on the method outlined in R9.7 can be employed.

```
function nvar = noisepwr1(num,den)
% Computes the output noise variance due
% to input quantization of a digital filter
% based on a partial-fraction approach
%
% num and den are the numerator and denominator
% polynomial coefficients of the IIR transfer function
%
[r,p,K] = residue(num,den);
R = size(r,1);
R2 = size(K,1);
if R2 > 1
    disp('Cannot continue...');
    return;
end
if R2 == 1
    nvar = K^2;
    else
    nvar = 0;
end
% Compute round off noise variance
for k = 1:R,
    for m = 1:R,
        integral = r(k)*conj(r(m))/(1-p(k)*conj(p(m)));
        nvar = nvar + integral;
    end
end
disp('Output Noise Variance = ');disp(real(nvar))
```

Questions:

Q9.16 Write a MATLAB program using the function noisepwr1 to compute the normalized output roundoff noise variance of a fourth order elliptic lowpass filter with the following specifications: a passband ripple of 0.5 dB, a minimum stopband attenuation of 50 dB, and a passband edge at 0.45. Run this program and determine the normalized output roundoff noise variance of this filter.

Q9.17 Repeat Question Q9.16 for a sixth order Type 2 Chebyshev bandpass filter with stopband edges at 0.3 and 0.75, and a minimum stopband attenuation of 60 dB.

Often it is convenient to compute an approximate value of the normalized output noise variance using Eq. (9.13) (see R9.8). The function noisepwr2 given below is based on this approach.

```
function nvar = noisepwr2(num,den)
% Computes the approximate output noise variance due
```

```
% to input quantization of a digital filter
% by summing the square of the impulse response samples
%
% num and den are the numerator and denominator
% polynomial coefficients of the IIR transfer function
%
x = 1;
order = max(length(num),length(den))-1;
si = [zeros(1,order)];
nvar = 0; k = 1;
while k > 0.0000001
    [y,sf] = filter(num,den,x,si);
    si = sf; k = abs(y)*abs(y);
    nvar = nvar + k;
    x = 0;
end
disp('Output Noise Variance = ');disp(nvar)
```

Question:

Q9.18   Write a MATLAB program using the function `noisepwr2` to compute the normalized output roundoff noise variance of an LTI causal stable digital filter. Using this program determine the normalized output noise variance of the filter of Question Q9.16 and compare the result with that obtained in Question Q9.16.

9.7   Analysis of Arithmetic Round-off Errors

The output noise variance due to the round-off of all products can be easily carried out using MATLAB. However, the output round-off noise always should be computed only after the digital filter structure has been scaled as the scaling process may introduce additional multipliers in the system. In practice, most of the scaling multipliers can be absorbed into the existing feed-forward multipliers without a significant increase in the total number of multipliers and, hence, noise sources. For the scaled structure, the expression for the output round-off noise of Eq. (9.19) thus changes to

$$\sigma_\gamma^2 = \sigma_o^2 \sum_{\ell=1}^{L} \bar{k}_\ell \left(\frac{1}{2\pi j} \oint_C \bar{G}_\ell(z)\, \bar{G}_\ell(z^{-1})\, z^{-1}\, dz \right), \tag{9.29}$$

where $\bar{k}_\ell$ is the total number of multipliers feeding the ℓth adder with $\bar{k}_\ell \geq k_\ell$ and $\bar{G}_\ell(z)$ is the modified noise transfer function from the input of the ℓth adder to the filter output.

The dynamic range scaling using the $\mathcal{L}_2$-norm rule can be easily performed using MATLAB by actual simulation of the digital filter structure. If we denote the impulse response from the input of the digital filter to the output of the r-th branch node as $\{f_r[n]\}$ and assume that the branch nodes have been ordered in accordance to their precedence relations with increasing r [Cro75], then we can compute the $\mathcal{L}_2$-norm of $\{f_1[n]\}$ first and then scale the input by a multiplier $k_1 = 1/\|F_1\|_2$. Next, we compute the $\mathcal{L}_2$-norm of $\{f_2[n]\}$ and scale the multipliers feeding into the second adder by dividing with a constant $k_2 = 1/\|F_2\|_2$. This process is continued until the output node has been scaled to yield an $\mathcal{L}_2$-norm of unity.

Project 9.5 Cascade Form IIR Digital Filter Structure

In this project you will learn the scaling of an IIR digital filter realized in cascade form using the $\mathcal{L}_2$-norm rule and the computation of the total outout noise variance due to all product round-offs of the scaled structure.

Question:

Q9.19 The filter to be designed is a lowpass ellptic filter with the following specifications: a passband edge at 0.25, a stopband edge at 0.5, a passband ripple of 0.5 dB, and a minimum stopband attenuation of 50 dB. Write a MATLAB program to determine the numerator and denominator polynomial coefficients of the individual second-order sections. You will need to use the following MATLAB functions: `ellipord`, `ellip`, and `zp2sos`.

A scaled cascade realization of the above transfer function with each section in Direct Form II is shown in Figure 9.7.

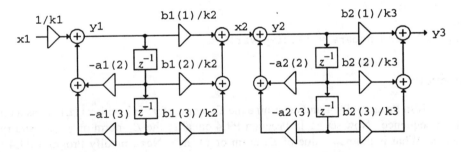

Figure 9.7 Cascade realization of the transfer function generated by Question Q9.19 in Direct Form II.

Question

Q9.20 What are the values of the constants `b1(1)`, `b1(2)`, `b1(3)`, `b2(1)`, `b2(2)`, `b2(3)`, `a1(2)`, `a1(3)`, `a2(2)`, and `a2(3)`?

The MATLAB program simulating this structure is given by Program P9_4 below. [2]

```
% Program P9_4
% Scaling and Round-off Noise Calculation
% of the Cascade-Form Structure of Figure 9.6
%
```

---

[2]The first coefficient in the denominator coefficients should be a 1.

```
b1 = input('Numerator coeffs.  of Sec. 1 = ');
a1 = input('Denominator coeffs.  of Sec. 1 = ');
b2 = input('Numerator coeffs.  of Sec. 2 = ');
a2 = input('Denominator coeffs.  of Sec. 2 = ');
format long
k1 = 1;
k2 = 1;
k3 = 1;
x1 = 1/k1;
si1 = [0 0]; si2 = [0 0];
var = 0; K = 1;
while K > 0.0000001
    y1 = - a1(2)*si1(1) - a1(3)*si1(2) + x1;
    x2 = (b1(1)*y1 + b1(2)*si1(1)+ b1(3)*si1(2))/k2;
    y2 = - a2(2)*si2(1) - a2(3)*si2(2) + x2;
    y3 = (b2(1)*y1 + b2(2)*si1(1)+ b2(3)*si1(2))/k3;
    si1(2) = si1(1); si1(1) = y1;
    si2(2) = si2(1); si2(1) = y2;
% Approximate L2 norm square computation
    K = abs(y1)*abs(y1);
    var = var + K;
    x1 = 0;
end
disp('L2 norm square = ');disp(var);
```

Question:

Q9.21 Run Program P9_4 and determine the $\mathcal{L}_2$-norm of y1[n] with x1[n] set as a unit sample sequence. Now set k1 in Program P9_4 equal to the $\mathcal{L}_2$-norm of y1[n] and run it again. What is the new value of $\mathcal{L}_2$-norm of y1[n]? Next, modify Program P9_4 to compute the $\mathcal{L}_2$-norm of y2[n] and run the modified program. Then set k2 in Program P9_4 equal to the $\mathcal{L}_2$-norm of y2[n] and run it again. What is the new value of the $\mathcal{L}_2$-norm of y2[n]? Finally, modify Program P9_4 to compute the $\mathcal{L}_2$-norm of y3[n] and run the modified program. Then set k3 in Program P9_4 equal to the $\mathcal{L}_2$-norm of y3[n] and run it again. What is the new value of the $\mathcal{L}_2$-norm of y3[n]?

The Program P9_4 with slight modifications can also be used to determine the total output product round-off noise variance of the structure of Figure 9.7.

Questions:

Q9.22 If all products are assumed to be quantized before addition, in the noise model of the structure of Figure 9.7 how many noise sources are entering the adder with output y1[n], how many noise sources are entering the adder with output y2[n], and how many noise sources are entering the adder with output y3[n]?

Q9.23 To determine the output product round-off noise variance due to a single noise source into the adder with output y1[n], replace the statement "x1 = 1/k1;" in the modified Program P9_4 (keeping all current values of the scaling constants) with the statement "x1 = 1;" and run it. To determine the output product round-off noise variance due to a single noise source into the adder with output y2[n], replace the statement "x1 = 1;" in the modified Program P9_4 (keeping all current values of the scaling constants) with the statement "x2 = 1;", the statement "x1 = 0;" with the statement "x2 = 0;", and run it. What are the values of the output product roundoff noise variance for the case if all products are quantized before additions, and if all products are quantized after additions.

Q9.24 Interchange Sections 1 and 2 in Figure 9.7, and repeat Questions Q9.22 and Q9.23 with the values of the vectors a1, b1, a2, and b2 appropriately changed. Which structure has the lowest output product round-off noise variance?

9.8 Low-Sensitivity Digital Filters

Project 9.6   Low Passband Sensitivity IIR Filters

In this project you will investigate using MATLAB, the low passband sensitivity property of the parallel allpass structure of Figure 6.13, realizing a bounded-real IIR transfer function.

Question:

Q9.25 Using the functions elliptord and ellip design an odd-order lowpass elliptic transfer function with the following specifications: passband edge at 0.4, stopband edge at 0.55, passband ripple of 0.5 dB, and minimum stopband attenuation of 50 dB. Develop the parallel allpass realization using the pole-interlacing property (see R6.19). How many multipliers are required in this realization? Using MATLAB, develop the gain responses of the transfer function of the parallel allpass realization of the above lowpass filter with unquantized and quantized coefficients (with six bits for the fractional part of the binary equivalent), and plot both responses on the same figure. Does the parallel allpass structure exhibit low passband sensitivity?

Project 9.7   Low Passband Sensitivity FIR Filters

In this project you will investigate using MATLAB, the low passband sensitivity property of the delay-complementary structure of Figure 9.6 realizing a bounded-real Type 1 FIR transfer function.

Questions:

Q9.26 Using the function remez design a linear-phase FIR filter of length 15 with a normalized passband edge at 0.55, a normalized stopband edge at 0.65, and equal weights to passband and stopband ripples. Using the function freqz compute the magnitude response of the FIR filter at 1024 equally spaced frequency points between 0 and 1, and divide the impulse response samples of the filter by the maximum of the absolute value of the magnitude response obtained using the command max. The scaled transfer function $H(z)$ should be a BR function. Verify this property by computing the frequency response of the scaled filter $H(z)$. Quantize the coefficients of the scaled filter $H(z)$ using the function a2dT with five bits assigned to the fractional part of the binary equivalent. Plot the magnitude responses of the scaled filter $H(z)$ and the scaled quantized filter on the same figure, and comment on your results.

Q9.27   Determine the coefficients of the delay-complementary filter $G(z)$ of the scaled FIR filter developed in Question Q9.26. Determine the the the zeros of the delay-complementary filter $G(z)$ using the command roots and form the factor $G_b(z)$ composed of the zeros on the unit circle. Determine the remaining factor $G_a(z)$ of $G(z)$ by deconvolving $G(z)$ with $G_b(z)$ using the command deconv. Quantize the coefficients of $G_a(z)$ and individual factors of $G_b(z)$ using the function a2dT with five bits assigned to the fractional part of the binary equivalents. Develop the delay-complementary filter of $G(z)$ realized as a cascade of $G_a(z)$ and $G_b(z)$ with quantized coefficients and plot its magnitude response. Does this realization of $H(z)$ exhibit low passband sensitivity?

9.9   Limit Cycles

Project 9.8   Granular Limit Cycle Generation

In this project you will investigate the generation of granular limit cycles in a first-order IIR filter given by

$$\hat{y}[n] = Q\left(\alpha\,\hat{y}[n-1]\right) + x[n], \tag{9.30}$$

where $\hat{y}[n]$ is the output obtained by rounding the product $\alpha\,\hat{y}[n-1]$ and $x[n]$ is the input. To this end Program P9_5 given below can be used.

```
% Program P9_5
% Granular Limit Cycles in First-Order IIR Filter
%
alpha = input('Type in the value of alpha = ');
yi = 0; x = 0.04;
for n = 1:21
    y(n) = a2dR(alpha*yi,5) + x;
    yi = y(n); x = 0;
end
k = 0:20;
stem(k,y)
ylabel('Amplitude'); xlabel('Time index n')
```

Questions:

Q9.28   What is the purpose of the function a2dR? How many bits are assigned to the fractional part of the binary equivalent? Run the above program for the following values of α: -0.55 and 0.55. Comment on your results.

Q9.29 Repeat Question Q9.28 varying the number of bits assigned to the fractional part of the binary equivalent.

Q9.30 Repeat Question Q9.28 with different values for the filter coefficient α. Comment on your results.

Project 9.9 Overflow Limit Cycle Generation

In this project you will investigate the generation of overflow limit cycles in a second-order IIR filter given by

$$\hat{y}[n] = \mathcal{Q}\left(-\alpha_1 \hat{y}[n-1] - \alpha_2 \hat{y}[n-2]\right) + x[n], \tag{9.31}$$

where $\hat{y}[n]$ is the output obtained by rounding the sum of the products $-\alpha_1 \hat{y}[n-1]$ and $-\alpha_2 \hat{y}[n-2]$, and $x[n]$ is the input. To this end Program P9_6 given below can be used.

```
% Program P9_6
% Overflow Limit Cycles in Second-Order IIR Filter
%
alpha = input('\alpha_{1} and \alpha_{2} values = ');
yi1 = 0.75; yi2 = -0.75;
for n = 1:41
    y(n) = - alpha(1)*yi1 - alpha(2)*yi2;
    y(n) = a2dR(y(n),3);
    yi2 = yi1; yi1 = y(n);
end
k = 0:40;
stem(k,y)
ylabel('Amplitude'); xlabel('Time index n')
```

Questions:

Q9.31 How many bits are assigned to the fractional part of the binary equivalent? Run the above program for the following values of the filter coefficients: $\alpha_1 = -0.875$ and $\alpha_2 = 0.875$. Comment on your results. How does this limit cycle differ from the limit cycle generated in Question Q9.28?

Q9.32 Repeat Question Q9.31 varying the number of bits assigned to the fractional part of the binary equivalent.

Q9.33 Repeat Question Q9.31 with different values for the filter coefficients α_1 and α_2. Comment on your results.

Q9.34 Repeat Question Q9.31 by replacing the function a2dR with a2dT in Program P9_6, which quantizes the sum of the products of $-\alpha_1 \hat{y}[n-1]$ and $-\alpha_2 \hat{y}[n-2]$ using truncation. Comment on your results.

9.10 Background Reading

[1] A. Antoniou. *Digital Filters: Analysis, Design, and Applications*. McGraw-Hill, New York NY, second edition, 1993. Ch. 11.

[2] E. Cunningham. *Digital Filtering: An Introduction*. Houghton-Mifflin, Boston MA, 1992. Ch. 8.

[3] D.J. DeFatta, J.G. Lucas, and W.S. Hodgkiss. *Digital Signal Processing: A System Design Approach*. Wiley, New York NY, 1988. Ch. 9.

[4] L.B. Jackson. *Digital Filters and Signal Processing*. Kluwer, Boston MA, third edition, 1996. Ch. 11.

[5] R. Kuc. *Introduction to Digital Signal Processing*. McGraw-Hill, New York NY, 1988. Ch. 10.

[6] S.K. Mitra. *Digital Signal Processing: A Computer-Based Approach*. McGraw-Hill, New York NY, 1998. Sec. 8.4, Ch. 9.

[7] A.V. Oppenheim and R.W. Schafer. *Discrete-Time Signal Processing*. Prentice-Hall, Englewood Cliffs NJ, 1989, Secs. 6.7 – 6.10.

[8] S.J. Orfanidis. *Introduction to Signal Processing*. Prentice-Hall, Englewood Cliffs NJ, 1996, Ch. 2, Sec. 7.6.

[9] B. Porat. *A Course in Digital Signal Procesing*. Wiley, New York NY, 1996. Sec. 11.4 – 11.8.

[10] J.G. Proakis and D.G. Manolakis. *Digital Signal Processing: Principles, Algorithms, and Applications*. Prentice-Hall, Englewood Cliffs NJ, third edition, 1996. Sec. 1.4, Secs. 7.5 – 7.7.

[11] R.A. Roberts and C.T. Mullis. *Digital Signal Processing*. Addison-Wesley, Reading MA, 1987. Ch. 9.

Multirate Digital Signal Processing 10

10.1 Introduction

The digital signal processing structures discussed so far belong to the class of single-rate systems as the sampling rates at the input and the output and all internal nodes are the same. There are applications where it is necessary and often convenient to have unequal rates of sampling at various parts of the system including the input and the output. In this laboratory exercise you will investigate first using MATLAB the properties of the up-sampler and the down-sampler, the two basic components of a multirate system. You will then investigate their use in designing more complex systems, such as interpolators and decimators, and filter banks.

10.2 Background Review

R10.1 The up-sampler, shown in Figure 10.1, is employed in the increase of the sampling rate of a sequence $x[n]$ by an integer factor $L > 1$. It inserts $L-1$ equidistant zero-valued samples between each consecutive pair of samples of $x[n]$ to develop an output sequence $x_u[n]$ given by

$$x_u[n] = \begin{cases} x[n/L], & n = 0, \pm L, \pm 2L, \ldots, \\ 0, & \text{otherwise.} \end{cases} \tag{10.1}$$

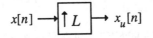

Figure 10.1 Represenation of an up-sampler.

The up-sampler is a linear but time-varying discrete-time system. In the z-domain, its input-output relation is given by

$$X_u(z) = X(z^L), \tag{10.2}$$

where $X(z)$ and $X_u(z)$ denote, respectively, the z-transforms of $x[n]$ and $x_u[n]$. For $z = e^{j\omega}$ the above equation becomes $X_u(e^{j\omega}) = X(e^{j\omega L})$.

R10.2 The down-sampler, shown in Figure 10.2, is employed in the decrease of the sampling rate of a sequence $x[n]$ by an integer factor $M > 1$. It removes $M-1$ in-between samples to generate an output sequence $y[n]$ according to the relation

$$y[n] = x[nM].$$ (10.3)

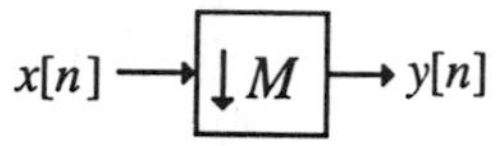

$$x[n] \longrightarrow \boxed{\downarrow M} \longrightarrow y[n]$$

Figure 10.2 Representation of a down-sampler.

The down-sampler is a linear but time-varying discrete-time system. In the z-domain, its input-output relation is given by

$$Y(z) = \frac{1}{M} \sum_{k=0}^{M-1} X(z^{1/M} W_M^{-k}),$$ (10.4)

where $X(z)$ and $Y(z)$ denote, respectively, the z-transforms of $x[n]$ and $y[n]$. For $z = e^{j\omega}$ the above equation becomes

$$Y(e^{j\omega}) = \frac{1}{M} \sum_{k=0}^{M-1} X(e^{j(\frac{\omega - 2\pi k}{M})}),$$ (10.5)

It follows from Eq. (10.5) that the aliasing due to a factor-of-M down-sampling is absent if and only if the signal $x[n]$ is bandlimited to $\pm \pi / M$.

R10.3   A cascade of a factor-of-M down-sampler with a factor-of-L up-sampler (Figure 10.3) is *interchangeable* with no change in the input-output relation if and only if M and L are *relatively prime*, that is, M and L do not have a common factor that is an integer $k > 1$.

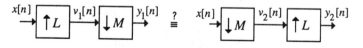

Figure 10.3   An up-sampler in cascade with a down-sampler.

R10.4   Two other cascade equivalences are shown in Figure 10.4.

R10.5   The unwanted images in the spectra of the up-sampled signal $x_u[n]$, due to the periodic repetition of the spectrum of $x[n]$ caused by the up-sampler, is removed by a low-pass filter $H(z)$ as indicated in Figure 10.5(a) after the up-sampling operation. The system of Figure 10.5(a) is called an *interpolator*. The specifications for the lowpass *interpolation filter* $H(z)$ are given by

$$|H(e^{j\omega})| = \begin{cases} L, & |\omega| \le \omega_c / L, \\ 0, & \pi / L \le |\omega| \le \pi. \end{cases}$$ (10.6)

where ω_c denotes the highest frequency that need to be preserved in the signal to be interpolated.

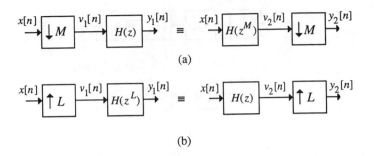

Figure 10.4 Cascade equivalences.

R10.6 Similarly, prior to down-sampling the signal $v[n]$ should be band limited to $|\omega| < \pi/M$ by means of a lowpass filter $H(z)$ to avoid aliasing caused by the down-sampling operation. The system of Figure 10.5(b) is called a *decimator*. The specifications for the lowpass *decimation filter* $H(z)$ are given by

$$|H(e^{j\omega})| = \begin{cases} 1, & |\omega| \leq \omega_c/M, \\ 0, & \pi/M \leq |\omega| \leq \pi. \end{cases} \tag{10.7}$$

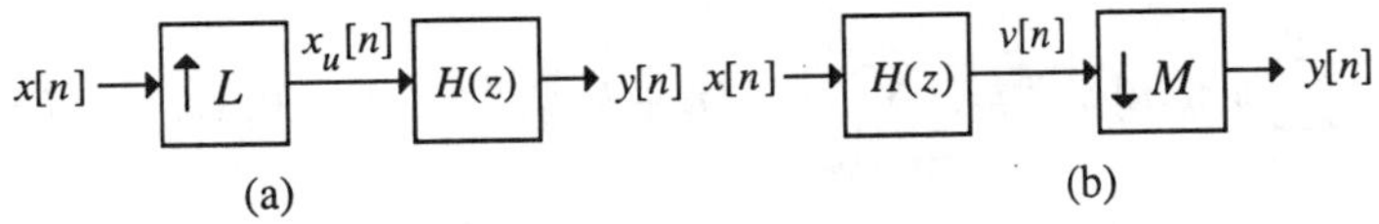

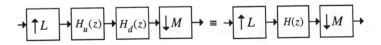

Figure 10.5 (a) Interpolator and (b) decimator.

R10.7 The desired configuration for a fractional sampling rate alteration is as indicated in Figure 10.6 where the lowpass filter $H(z)$ has a normalized stopband cutoff frequency at

$$\omega_s = \min\left(\frac{\pi}{L}, \frac{\pi}{M}\right). \tag{10.8}$$

Figure 10.6 Fractional rate sampling rate alteration scheme.

R10.8 In general the computational efficiency is improved significantly by designing the sampling rate alteration system as a casacde of several stages. Figure 10.7(a) illustrates a two-stage design of a factor-of-L interpolator where $L = L_1 L_2$. Likewise, Figure 10.7(b) illustrates a two-stage design of a factor-of-M decimator where $M = M_1 M_2$.

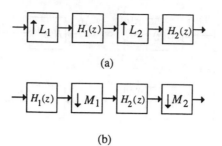

(a)

(b)

Figure 10.7   Two-stage implementations: (a) interpolator, and (b) decimator.

R10.9   The M-band *polyphase decomposition* of the z-transform $X(z)$ of an arbitrary sequence $\{x[n]\}$

$$X(z) = \sum_{n=-\infty}^{\infty} x[n]\, z^{-n}, \tag{10.9}$$

is given by

$$X(z) = \sum_{k=0}^{M-1} z^{-k}\, X_k(z^M), \tag{10.10}$$

where

$$X_k(z) = \sum_{n=-\infty}^{\infty} x_k[n]\, z^{-n} = \sum_{n=-\infty}^{\infty} x[Mn+k]\, z^{-n}, \quad k = 0, 1, \ldots, M-1. \tag{10.11}$$

The subsequences $\{x_k[n]\}$ are called the *polyphase components* of the parent sequence $\{x[n]\}$, and the functions $X_k(z)$, given by the z-transform of $\{x_k[n]\}$, are called the *polyphase components* of $X(z)$. The relation between the subsequences $\{x_k[n]\}$ and the original sequence $\{x[n]\}$ is given by

$$x_k[n] = x[Mn+k], \quad k = 0, 1, \ldots, M-1. \tag{10.12}$$

R10.10   The *Type I polyphase decomposition* of an Nth order FIR transfer function $H(z)$ is given by

$$H(z) = \sum_{k=0}^{M-1} z^{-k}\, E_k(z^M) \tag{10.13}$$

whereas its *Type II polyphase decomposition* is given by

$$H(z) = \sum_{\ell=0}^{M-1} z^{-(M-1-\ell)}\, R_\ell(z^M) \tag{10.14}$$

where

$$R_\ell(z) = E_{M-1-\ell}(z), \quad \ell = 0, 1, \ldots, M-1. \tag{10.15}$$

The direct realization of $H(z)$ based on the decompositions of Eqs. (10.13) and (10.14) is shown in Figure 10.8.

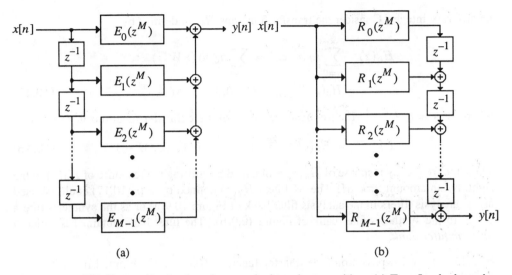

Figure 10.8 FIR filter realization based on a polyphase decomposition: (a) Type I polyphase decomposition and (b) Type II polyphase decomposition.

R10.11 The digital filter bank is a set of digital bandpass filters with either a common input or a summed output as shown in Figure 10.9. The structure of Figure 10.9(a) is called an M-band *analysis filter bank* with the subfilters $H_k(z)$ known as the *analysis filters*. Likewise, the structure of Figure 10.9(b) is called an L-band *synthesis filter bank* with the subfilters $F_k(z)$ known as the *synthesis filters*.

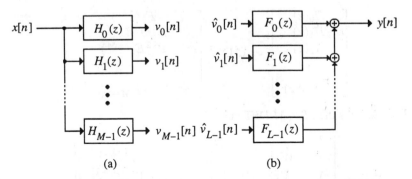

Figure 10.9 (a) Analysis filter bank and (b) synthesis filter bank.

R10.12 Let $H_0(z)$ represent a causal lowpass digital filter with an impulse response $h_0[n]$:

$$H_0(z) = \sum_{n=0}^{\infty} h_0[n]\, z^{-n},\qquad (10.16)$$

with a passband edge ω_p, and a stopband edge ω_s, where $\omega_p < \pi/M < \omega_s$ with M being an arbitrary integer. Consider the transfer functions $H_k(z)$ defined by

$$H_k(z) = \sum_{n=0}^{\infty} h_k[n]\, z^{-n} = \sum_{n=0}^{\infty} h_0[n]\, (z\, W_M^k)^{-n}$$

$$= H_0(z\, W_M^k), \qquad k = 0, 1, \ldots, M-1, \tag{10.17}$$

where $W_M = e^{-j2\pi/M}$. The frequency response of these transfer functions is given by

$$H_k(e^{j\omega}) = H_0(e^{j(\omega - \frac{2\pi k}{M})}), \qquad k = 0, 1, \ldots, M-1. \tag{10.18}$$

Thus, the frequency response of $H_k(z)$ is obtained by shifting the response of $H_0(z)$ to the right, by an amount $2\pi k/M$. The M filters $H_k(z)$ defined by Eq. (10.17) could be used as the analysis filters in the analysis filter bank of Figure 10.9(a) or as the synthesis filters $F_k(z)$ in the synthesis filter bank of Figure 10.9(b). The filter bank obtained is called a *uniform filter bank*.

R10.13   If the lowpass prototype transfer function $H_0(z)$ is represented in its M-band polyphase form:

$$H(z) = \sum_{\ell=0}^{M-1} z^{-\ell}\, E_\ell(z^M), \tag{10.19}$$

where $E_\ell(z)$ is the ℓth polyphase component of $H_0(z)$ given by

$$E_\ell(z) = \sum_{n=0}^{\infty} e_\ell[n]\, z^{-n} = \sum_{n=0}^{\infty} h_0[\ell + nM]\, z^{-n}, \qquad 0 \le \ell \le M-1, \tag{10.20}$$

the M filters, $H_k(z),\ k = 0, 1, \ldots, M-1$, can be expressed in matrix form as

$$\begin{bmatrix} H_0(z) \\ H_1(z) \\ H_2(z) \\ \vdots \\ H_{M-1}(z) \end{bmatrix} = M\, \mathbf{D}_M^{-1} \begin{bmatrix} E_0(z^M) \\ z^{-1} E_1(z^M) \\ z^{-2} E_2(z^M) \\ \vdots \\ z^{-(M-1)} E_{M-1}(z^M) \end{bmatrix}, \tag{10.21}$$

where $\mathbf{D}_M$ denotes the $M \times M$ DFT matrix:

$$\mathbf{D}_M = \begin{bmatrix} 1 & 1 & 1 & \cdots & 1 \\ 1 & W_M^1 & W_M^2 & \cdots & W_M^{M-1} \\ 1 & W_M^2 & W_M^4 & \cdots & W_M^{2(M-1)} \\ \vdots & \vdots & \vdots & \ddots & \vdots \\ 1 & W_M^{(M-1)} & W_M^{2(M-1)1} & \cdots & W_M^{(M-1)^2} \end{bmatrix}. \tag{10.22}$$

An efficient implementation of the M-band analysis filter bank based on Eq. (10.21) is thus as shown in Figure 10.10 where the prototype lowpass filter $H_0(z)$ has been implemented in a polyphase form. The structure of Figure 10.10 is more commonly known as the *uniform DFT analysis filter bank*. The corresponding polyphase implementation of a *uniform DFT synthesis filter bank* can be similarly derived and is shown in Figure 10.11.

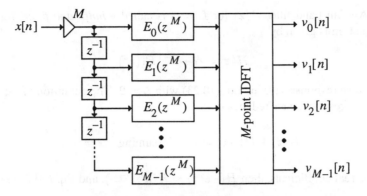

Figure 10.10 Uniform DFT analysis filter bank.

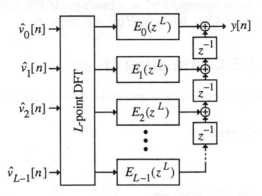

Figure 10.11 Uniform DFT synthesis filter bank.

R10.14 A lowpass filter $H(z)$ with a passband edge ω_p and a stopband edge ω_s satisfying $\omega_p < \pi/M < \omega_s$ is called an *Lth band filter* or a *Nyquist filter* if its impulse response $h[n]$ also satisfies the condition

$$h[Ln] = \begin{cases} \alpha, & n = 0, \\ 0, & \text{otherwise.} \end{cases} \qquad (10.23)$$

For such a filter, the polyphase decomposition is of the form

$$H(z) = \alpha + \sum_{\ell=1}^{L-1} z^{-\ell} E_\ell(z^L), \qquad (10.24)$$

and

$$\sum_{k=0}^{L-1} H(z\,W_L^k) = L\alpha = 1, \quad (\text{assuming} \quad \alpha = \frac{1}{L}). \qquad (10.25)$$

R10.15   An Lth band filter $H(z)$ for $L = 2$ is called a *half-band filter*, for which the transfer function is given by

$$H(z) = \alpha + z^{-1}E_1(z^2), \tag{10.26}$$

and its impulse response satisfies Eq. (10.23) with $L = 2$. The condition of Eq. (10.25) on the frequency response here reduces to

$$H(z) + H(-z) = 1 \quad \text{(assuming } \alpha = \frac{1}{2}). \tag{10.27}$$

If $H(z)$ has real coefficients, then $H(-e^{j\omega}) = H(e^{j(\pi-\omega)})$, and Eq. (10.27) leads to

$$H(e^{j\omega}) + H(e^{j(\pi-\omega)}) = 1. \tag{10.28}$$

The length of the impulse response $h[n]$ of a linear-phase FIR half-band filter $H(z)$ is restricted to be of the form $4K + 3$ (unless $H(z)$ is a constant).

10.3   MATLAB Commands Used

The MATLAB commands you will encounter in this exercise are as follows:

General Purpose Commands

    length

Operators and Special Characters

    :   .   +   –   *   /   ;
    %   .*   =

Language Constructs and Debugging

    end   for

Elementary Matrices and Matrix Manipulation

    pi   sum   :

Elementary Functions

    abs   exp   log10   sin

Two-Dimensional Graphics

    axis   clf   grid   plot   stem
    subplot   title   xlabel   ylabel

Signal Processing Toolbox

decimate	fir2	freqz	hamming	interp
remez	resample	sinc		

For additional information on these commands, see the *MATLAB Reference Guide* [Mat94] and the *Signal Processing Toolbox User's Guide* [Mat96] or type help commandname in the Command window. A brief explanation of the MATLAB functions used here can be found in Appendix B.

10.4   Basic Sampling Rate Alteration Devices

The objective of this section is to investigate using MATLAB the operations of the up-sampler and the down-sampler both in the time-domain and in the frequency-domain.

Project 10.1   Input-Output Relations in the Time-Domain

Program P10_1 can be used to study the operation of a up-sampler.

```
% Program 10_1
% Illustration of Up-Sampling by an Integer Factor
%
clf;
n = 0:50;
x = sin(2*pi*0.12*n);
y = zeros(1, 3*length(x));
y([1:  3:  length(y)]) = x;
subplot(2,1,1)
stem(n,x);
title('Input Sequence');
xlabel('Time index n');ylabel('Amplitude');
subplot(2,1,2)
stem(n,y(1:length(x)));
title('Output Sequence');
xlabel('Time index n');ylabel('Amplitude');
```

Questions:

Q10.1   What is the angular frequency in radians of the sinusoidal sequence? What is its length? What is the up-samping factor L?

Q10.2   How is the up-sampling operation implemented in Program P10_1?

Q10.3   Run Program P10_1 and verify that the relation between the output and the input sequences satisfies Eq. (10.1).

Q10.4   Repeat Question Q10.3 for two different values of the angular frequency and two different values of the up-samping factor L.

Q10.5   Modify Program P10_1 to study the operation of an up-sampler on a ramp sequence.

Program P10_2 can be used to study the operation of a down-sampler.

```
% Program P10_2
% Illustration of Down-Sampling by an Integer Factor
%
clf;
n = 0:  49;
m = 0:  50*3 - 1;
x = sin(2*pi*0.042*m);
y = x([1:  3:  length(x)]);
subplot(2,1,1)
stem(n, x(1:50)); axis([0 50 -1.2 1.2]);
title('Input Sequence');
xlabel('Time index n');
ylabel('Amplitude');
subplot(2,1,2)
stem(n, y); axis([0 50 -1.2 1.2]);
title('Output Sequence');
xlabel('Time index n');
ylabel('Amplitude');
```

Questions:

Q10.6   What is the angular frequency in radians of the sinusoidal sequence? What is its length? What is the down-samping factor M?

Q10.7   How is the down-sampling operation implemented in Program P10_2?

Q10.8   Run Program P10_2 and verify that the relation between the output and the input sequences satisfies Eq. (10.3).

Q10.9   Repeat Question Q10.8 for two different values of the angular frequency and two different values of the down-samping factor M.

Project 10.2   Input-Output Relations in the Frequency-Domain

To demonstrate the effect of up-sampling and down-sampling in the frequency-domain we need to create a finite-length input sequence that is also band-limited in the frequency-domain. To this end we can utilize the M-function fir2 (see Section 7.5).

Program P10_3 can be employed to study the frequency-domain properties of the up-sampler.

```
% Program P10_3
% Effect of Up-sampling in the Frequency-Domain
% Use fir2 to create a band-limited input sequence
clf;
freq = [0 0.45 0.5 1];
mag = [0 1 0 0];
x = fir2(99, freq, mag);
% Evaluate and plot the input spectrum
[Xz, w] = freqz(x, 1, 512, 'whole');
subplot(2,1,1);
plot(w/pi, abs(Xz)); axis([0 1 0 1]); grid
xlabel('\omega/\pi'); ylabel('Magnitude');
title('Input Spectrum');
subplot(2,1,2);
% Generate the up-sampled sequence
L = input('Type in the up-sampling factor = ');
y = zeros(1, L*length(x));
y([1:  L: length(y)]) = x;
% Evaluate and plot the output spectrum
[Yz, w] = freqz(y, 1, 512, 'whole');
plot(w/pi, abs(Yz)); axis([0 1 0 1]); grid
xlabel('\omega/\pi'); ylabel('Magnitude');
title('Output Spectrum');
```

Questions:

Q10.10   What is the length of the input sequence? What is its magnitude spectrum?

Q10.11   Run Program P10_3 for the following values of the up-sampling factor: $L = 2$, 3, and 5. Comment on the plots generated by the program. Do the results agree with Eq. (10.2)?

Program P10_4 can be employed to study the frequency-domain properties of the down-sampler.

```
% Program P10_4
% Effect of Down-sampling in the Frequency-Domain
% Use fir2 to create a band-limited input sequence
clf;
freq = [0 0.42 0.48 1]; mag = [0 1 0 0];
x = fir2(101, freq, mag);
% Evaluate and plot the input spectrum
[Xz, w] = freqz(x, 1, 512);
subplot(2,1,1);
plot(w/pi, abs(Xz)); grid
xlabel('\omega/\pi'); ylabel('Magnitude');
title('Input Spectrum');
```

```
% Generate the down-sampled sequence
M = input('Type in the down-sampling factor = ');
y = x([1:  M: length(x)]);
% Evaluate and plot the output spectrum
[Yz, w] = freqz(y, 1, 512);
subplot(2,1,2);
plot(w/pi, abs(Yz)); grid
xlabel('\omega/\pi'); ylabel('Magnitude');
title('Output Spectrum');
```

Questions:

Q10.12   What is the length of the input sequence? What is its magnitude spectrum?

Q10.13   Run Program P10_4 for the following values of the down-sampling factor: $M = 2, 3,$ and 5. Comment on the plots generated by the program. Do the results agree with Eq. (10.5)? What is the minimum value of M for which aliasing occurs?

10.5   Decimator and Interpolator Design and Implementation

The Signal Processing Toolbox includes three M-functions which can be employed to design and implement an interpolator or a decimator. The three M-functions are decimate, interp, and resample. Each function is available with several options. In this section you will study the decimation and interpolation operation using these functions.

Project 10.3   Decimator Design and Implementation

Program P10_5 illustrates the use of the M-function decimate in the design and implementation of a decimator with an integer-valued decimation factor M. In the option utilized in this program, decimate designs and uses a lowpass decimation filter with a stopband edge satisfying Eq. (10.7).

```
% Program P10_5
% Illustration of Decimation Process
%
clf;
M = input('Down-sampling factor = ');
n = 0:99;
x = sin(2*pi*0.043*n) + sin(2*pi*0.031*n);
y = decimate(x,M,'fir');
subplot(2,1,1);
stem(n,x(1:100));
title('Input Sequence');
xlabel('Time index n');ylabel('Amplitude');
```

```
subplot(2,1,2);
m = 0:(100/M)-1;
stem(m,y(1:100/M));
title('Output Sequence');
xlabel('Time index n');ylabel('Amplitude');
```

Questions:

Q10.14   What are the frequencies of the two sinusoidal sequences forming the input sequence? What is the length of the input?

Q10.15   What are the type and order of the decimation filter?

Q10.16   Run Program P10_5 for $M = 2$ and comment on the results.

Q10.17   Change the frequencies of the two sinusoidal sequences in the input signal to 0.045 and 0.029, and the length of the input to 120. Run the modified Program P10_5 for $M = 3$. Comment on your results.

Project 10.4   Interpolator Design and Implementation

Program P10_6 illustrates the use of the M-function `interp` in the design and implementation of an interpolator with an integer-valued interpolation factor L. `interp` designs and uses a lowpass interpolation filter with a stopband edge satisfying Eq. (10.6).

```
% Program P10_6
% Illustration of Interpolation Process
%
clf;
L = input('Up-sampling factor = ');
% Generate the input sequence
n = 0:49;
x = sin(2*pi*0.043*n) + sin(2*pi*0.031*n);
% Generate the interpolated output sequence
y = interp(x,L);
% Plot the input and the output sequences
subplot(2,1,1);
stem(n,x(1:50));
title('Input Sequence');
xlabel('Time index n'); ylabel('Amplitude');
subplot(2,1,2);
m = 0:(50*L)-1;
stem(m,y(1:50*L));
title('Output Sequence');
xlabel('Time index n'); ylabel('Amplitude');
```

Questions:

Q10.18 What are the frequencies of the two sinusoidal sequences forming the input sequence? What is the length of the input?

Q10.19 What are the type and order of the interpolation filter?

Q10.20 Run Program P10_6 for $L = 2$ and comment on the results.

Q10.21 Change the frequencies of the two sinusoidal sequences in the input signal to 0.045 and 0.029, and the length of the input to 40. Run the modified Program P10_6 for $L = 3$. Comment on your results.

Project 10.5   Fractional-Rate Sampling Rate Alteration

Program P10_7 illustrates the use of the M-function `resample` in the design and implementation of an interpolator with a fractional-rate interpolation factor L/M. `resample` designs and uses a lowpass interpolation filter with a stopband edge satisfying Eq. (10.8).

```
% Program 10_7
% Illustration of Sampling Rate Alteration by
% a Ratio of Two Integers
%
clf;
L = input('Up-sampling factor = ');
M = input('Down-sampling factor = ');
n = 0:29;
x = sin(2*pi*0.43*n) + sin(2*pi*0.31*n);
y = resample(x,L,M);
subplot(2,1,1);
stem(n,x(1:30));axis([0 29 -2.2 2.2]);
title('Input Sequence');
xlabel('Time index n'); ylabel('Amplitude');
subplot(2,1,2);
m = 0:(30*L/M)-1;
stem(m,y(1:30*L/M));axis([0 (30*L/M)-1 -2.2 2.2]);
title('Output Sequence');
xlabel('Time index n'); ylabel('Amplitude');
```

Questions:

Q10.22 What are the frequencies of the two sinusoidal sequences forming the input sequence? What is the length of the input?

Q10.23 What are the type and order of the band-limiting filter?

Q10.24 Run Program P10_7 for $L = 5$ and $M = 3$. Comment on the results.

Q10.25 Change the frequencies of the two sinusoidal sequences in the input signal to 0.045 and 0.029, and the length of the input to 40. Run the modified Program P10_7 for $L = 3$ and $M = 5$. Comment on your results.

10.6 Design of Filter Banks

Project 10.6 Design of Uniform Analysis/Synthesis Filter Banks

The design of an M-channel uniform analysis or synthesis filter bank can be carried out easily using Eq. (10.17) where $\{h_0[n]\}$ is a prototype lowpass filter with a passband edge ω_p and a stopband edge ω_s satisfying the condition $\omega_p < \pi/M < \omega_s$. The design can be carried out using Program P10_8.

```
% Program P10_8
% Design of Uniform DFT Filter Banks
clf;
% Design the prototype lowpass filter
b = remez(20, [0 0.2 0.25 1], [1 1 0 0], [10 1]);
w = 0:2*pi/255:2*pi; n = 0:20;
for k = 1:4;
    c = exp(2*pi*(k-1)*n*i/4);
    FB = b.*c;
    HB(k,:)  = freqz(FB,1,w);
end
% Plot magnitude responses of each filter
subplot(2,2,1)
plot(w/pi, abs(HB(1,:)));
xlabel('\omega/\pi');ylabel('Amplitude');
title('Filter No.1'); axis([0 2 0 1.1]);
subplot(2,2,2)
plot(w/pi,abs(HB(2,:)));
xlabel('\omega/\pi');ylabel('Amplitude');
title('Filter No.2');axis([0 2 0 1.1]);
subplot(2,2,3)
plot(w/pi,abs(HB(3,:)));
xlabel('\omega/\pi');ylabel('Amplitude');
title('Filter No.3'); axis([0 2 0 1.1]);
subplot(2,2,4)
plot(w/pi,abs(HB(4,:)));
xlabel('\omega/\pi');ylabel('Amplitude');
title('Filter No.4'); axis([0 2 0 1.1]);
```

Questions:

Q10.26 What are the type and order of the prototype lowpass filter in Program P10_8? What are its specifications? How many channels are in the filter bank?

Q10.27   Run Program P10_8 and comment on your results.

Q10.28   Modify Program P10_8 to design a three-channel uniform filter bank and plot their magnitude responses. Use the same prototype lowpass filter.

10.7   Design of Nyquist Filters

Project 10.7   Windowed Fourier Series Approach

A lowpass linear-phase Nyquist Lth band FIR filter with a cutoff at $\omega_c = \pi/L$ and a good frequency response can be readily designed via the windowed Fourier series approach described in Section R7.9. In this approach the impulse response coefficients of the lowpass filter are chosen as

$$h[n] = h_{LP}[n] \cdot w[n]. \tag{10.29}$$

where $h_{LP}[n]$ is the impulse response of an ideal lowpass filter with a cutoff at π/L and $w[n]$ is a suitable window function. If

$$h_{LP}[n] = 0 \quad \text{for } n = 0, \pm L, \pm 2L, \dots, \tag{10.30}$$

then Eq. (10.23) is indeed satisfied.

Now the impulse response $h_{LP}[n]$ of an ideal Lth band filter is obtained from Eq. (7.14) by substituting $\omega_c = \pi/L$ and is given by

$$h_{LP}[n] = \frac{\sin(\pi n/L)}{\pi n}, \quad -\infty \leq n \leq \infty. \tag{10.31}$$

It can be seen from the above that the impulse response coefficients do indeed satisfy the condition of Eq. (10.23). Hence, an Lth band filter can be designed by applying a suitable window function to Eq. (10.31).

The design of an Lth band lowpass filter can be carried out using Program P10_9.

```
% Program P10_9
% Design of Lth Band FIR Filter Using the
% Windowed Fourier Series Approach
%
clf;
K = 11;
n = -K:K;
% Generate the truncated impulse response of
% the ideal lowpass filter
b = sinc(n/2)/2;
% Generate the window sequence
win = hamming(23);
% Generate the coefficients of the windowed filter
fil = b.*win';
```

```
c = fil/sum(fil);
% Plot the gain response of the windowed filter
[h,w] = freqz(c,1,256);
g = 20*log10(abs(h));
plot(w/pi,g);axis([0 1 -90 10]); grid
xlabel('\omega/\pi');ylabel('Gain, dB');
```

Questions:

Q10.29 What are the value of L and the length of the Lth band FIR filter being designed by Program P10.9? What type of window is being used?

Q10.30 Run Program P10.9 and comment on your results. Print the coefficients of the Lth band filter and verify that it satisfies Eq. (10.23).

Q10.31 Repeat Question Q10.30 for a different value of L and a different window.

10.8 Background Reading

[1] R.E. Crochiere and L.R. Rabiner. *Multirate Digital Signal Processing*. Prentice-Hall, Englewood Cliffs NJ, 1996. Chs. 2, 4, 7.

[2] D.J. DeFatta, J.G. Lucas, and W.S. Hodgkiss. *Digital Signal Processing: A System Design Approach*. Wiley, New York NY, 1988. Ch. 7.

[3] S.K. Mitra. *Digital Signal Processing: A Computer-Based Approach*. McGraw-Hill, New York NY, 1998. Sec. 8.4 and Ch. 10.

[4] S.J. Orfanidis. *Introduction to Signal Processing*, Prentice-Hall, Englewood Cliffs, NJ, 1996, Ch. 12.

[5] B. Porat, *A Course in Digital Signal Procesing*. John Wiley, New York NY, 1996. Ch. 12.

[6] J.G. Proakis and D.G. Manolakis. *Digital Signal Processing: Principles, Algorithms, and Applications*. Prentice-Hall, Englewood Cliffs NJ, third Edition, 1996. Ch. 10.

[7] P.P. Vaidyanathan. *Multirate Systems and Filter Banks*. Prentice-Hall, Englewood Cliffs NJ, 1993. Chs. 4, 5.

Advanced Projects *11*

11.1 Introduction

This exercise contains 12 advanced projects for implementation using MATLAB. These projects make use of the concepts and tools of digital signal processing that you have learned in the previous 10 laboratory excercises. The projects considered here require careful planning before you start implementing them and may require reading additional materials beyond those to which you have been exposed so far. By completing these projects you will learn more about some interesting applications of the theory and algorithms of digital signal processing and gain a better appreciation of this field. In addition, you will be exposed to some recent advances in the field.

11.2 Discrete Transforms

Project 11.1 Subband Discrete Fourier Transform

The N-point DFT $X[k]$, $k = 0, 1, \cdots, N-1$, of a length-N sequence $x[n]$, $0 \le n \le N-1$, is given by the samples of its z-transform $X(z) = \sum_{n=0}^{N-1} x[n] \, z^{-n}$ evaluated on the unit circle at equally spaced points:

$$X[k] = X(z) \Big|_{z=W_N^{-k}}, \; = \sum_{n=0}^{N-1} x[n] \, W_N^{kn}, \quad 0 \le k \le N - 1, \tag{11.1}$$

where $W_N = e^{-j2\pi/N}$. The *subband discrete Fourier transform* (SB-DFT) computation method can be used to compute efficiently the approximate values of the dominant DFT samples in one or more portions of the frequency range of a sequence of a length that is a power of 2. To illustrate the basic idea behind this method, decompose $x[n]$ first into two length-$\frac{N}{2}$ subsequences $g_L[n]$ and $g_H[n]$ according to

$$g_L[n] = \frac{1}{2} \left(x[2n] + x[2n + 1] \right),$$
$$g_H[n] = \frac{1}{2} \left(x[2n] - x[2n + 1] \right), \quad 0 \le n \le \frac{N}{2} - 1. \tag{11.2}$$

Note that the original sequence $x[n]$ can be recovered from the above subsequences using

$$x[2n] = g_L[n] + g_H[n],$$
$$x[2n + 1] = g_L[n] - g_H[n], \quad 0 \le n \le \frac{N}{2} - 1. \tag{11.3}$$

The z-transform $X(z)$ of $x[n]$ can now be expressed as

$$X(z) = (1 + z^{-1})G_L(z^2) + (1 - z^{-1})G_H(z^2), \qquad (11.4)$$

where $G_L(z)$ and $G_H(z)$ are, respectively, the z-transforms of $g_L[n]$ and $g_H[n]$. The N-point DFT $X[k]$ of $x[n]$ thus can be expressed alternately as

$$X[k] = (1 + W_N^k)G_L[\langle k \rangle_{N/2}] + (1 - W_N^k)G_H[\langle k \rangle_{N/2}], \qquad 0 \le k \le N - 1, \quad (11.5)$$

where $G_L[k]$ and $G_H[k]$ are the $N/2$-point DFTs of the subsequences $g_L[n]$ and $g_H[n]$, respectively. The DFT computation of Eq. (11.5) has been called the *subband DFT* [She95]. Note that the two $N/2$-point DFTs can be computed using any FFT algorithms.

Equation (11.5) can be written in matrix form as

$$X[k] = \begin{bmatrix} 1 & W_N^k \end{bmatrix} \mathbf{R}_2^{-1} \begin{bmatrix} G_L[\langle k \rangle_{N/2}] \\ G_H[\langle k \rangle_{N/2}] \end{bmatrix}, \qquad (11.6)$$

where $\mathbf{R}_2 = \begin{bmatrix} 1 & 1 \\ 1 & -1 \end{bmatrix}$ is the 2×2 *Hadamard matrix*. The process can be repeated, resulting in a two-stage algorithm given by

$$X[k] = \begin{bmatrix} 1 & W_N^k & W_N^{2k} & W_N^{3k} \end{bmatrix} \mathbf{R}_4^{-1} \begin{bmatrix} G_{LL}[\langle k \rangle_{N/4}] \\ G_{LH}[\langle k \rangle_{N/4} \\ G_{HL}[\langle k \rangle_{N/4}] \\ G_{HH}[\langle k \rangle_{N/4} \end{bmatrix}, \qquad (11.7)$$

where $\mathbf{R}_4$ is the 4×4 Hadamard matrix and $G_{LL}[\langle k \rangle_{N/4}]$ and so forth are now $N/4$-point DFTs. This process can be continued until all the necessary DFTs are of length 2. If $N = 2^\mu$, the total number of complex multiplications required in a μ-stage subband DFT algorithm is the same as any Cooley-Tukey-type FFT algorithm, but requires more additions.

The DFT computation algorithm can be simplified by eliminating the calculations corresponding to the bands with negligible energy. For example, for a sequence $x[n]$ with mostly low frequency components, an approximate DFT is obtained by eliminating the contribution of $G_H[k]$ from Eq. (11.5):

$$X[k] \cong (1 + W_N^k)G_L[\langle k \rangle_{N/2}], \qquad 0 \le k \le \frac{N}{4} - 1, \qquad (11.8)$$

assuming $x[n]$ is a real-valued sequence. The high-frequency samples of $X[k]$ are assumed to be of zero value. The total computation requirements in using the approximate algorithm of Eq. (11.8) are now one-half of those required in the original algorithm of Eq. (11.5). Depending on the spectral property of the sequence $[n]$, in an M-stage decomposition, any one or more of the N/M dominant DFT samples can be approximately evaluated using the SB-DFT computation scheme.

Write a MATLAB program to determine the DFT of a sequence based on a two-stage SB-DFT computation scheme. Generate a length-64 real sequence with a band-limited spectrum restricted to the range $0 \leq \omega \leq \frac{\pi}{4}$ using the M-function `fir2`. Determine its approximate DFT using the SB-DFT program. Compare the low-frequency samples of the SB-DFT with those of the exact low frequency samples of the DFT obtained using the function `fft`. Using the command `flops`, compare also the computational complexities of both DFT computation schemes.

Next, modify the above program to determine the DFT of a sequence based on a 4-stage SB-DFT computation scheme. Determine the approximate DFT of the length-64 sequence generated earlier using the new SB-DFT program. Compare the low-frequency samples generated by the 4-stage SB-DFT with those generated by the 2-stage SB-DFT. Using the command `flops`, compare also the computational complexities of both SB-DFT computation schemes. Comment on your results.

Project 11.2 Nonuniform Discrete Fourier Transform

The N-point *nonuniform discrete Fourier transform* (NDFT) $X_{NDFT}[k]$ of a length-N sequence $x[n]$ is defined by [Bag98]

$$X_{NDFT}[k] = \sum_{n=0}^{N-1} x[n]\, z_k^{-n}, \qquad 0 \leq k \leq N-1, \qquad (11.9)$$

where $z_0, z_1, \cdots, z_{N-1}$ are N distinct points located arbitrarily in the z-plane. The above set of N equations can be expressed in matrix form as

$$
\begin{bmatrix}
X_{NDFT}[0] \\
X_{NDFT}[1] \\
\vdots \\
X_{NDFT}[N-1]
\end{bmatrix}
= \mathbf{D}_N
\begin{bmatrix}
x[0] \\
x[1] \\
\vdots \\
x[N-1]
\end{bmatrix}, \qquad (11.10)
$$

where

$$
\mathbf{D}_N =
\begin{bmatrix}
1 & z_0^{-1} & z_0^{-2} & \cdots & z_0^{-(N-1)} \\
1 & z_1^{-1} & z_1^{-2} & \cdots & z_1^{-(N-1)} \\
\vdots & \vdots & \vdots & \ddots & \vdots \\
1 & z_{N-1}^{-1} & z_{N-1}^{-2} & \cdots & z_{N-1}^{-(N-1)}
\end{bmatrix}, \qquad (11.11)
$$

is the $N \times N$ NDFT matrix. Note that for $z_k = e^{j2\pi k/N}$, $0 \leq k \leq N-1$, the NDFT reduces to the more conventional discrete Fourier transform (DFT) given in Eq. (11.1). The NDFT matrix of Eq. (11.11) is a *Vandermonde matrix*, and for distinct z_k it has an unique inverse. We can then solve Eq. (11.10) for the samples of the sequence $x[n]$ using

$$
\begin{bmatrix}
x[0] \\
x[1] \\
\vdots \\
x[N-1]
\end{bmatrix}
= \mathbf{D}_N^{-1}
\begin{bmatrix}
X_{NDFT}[0] \\
X_{NDFT}[1] \\
\vdots \\
X_{NDFT}[N-1]
\end{bmatrix}. \qquad (11.12)
$$

The inverse NDFT computation indicated above is not recommended in practice as the Vandermonde matrix is, in general, ill-conditioned and can lead to large numerical errors. The inverse NDFT can be computed more efficiently by evaluating the z-transform $X(z)$,

$$X(z) = \sum_{n=0}^{N-1} x[n]\, z^{-n}, \tag{11.13}$$

from the given N-point NDFT $X_{NDFT}[k]$ by using some type of polynomial interpolation method, such as the *Lagrange interpolation method*. In this method, $X(z)$ is expressed as

$$X(z) = \sum_{n=0}^{N-1} \frac{I_k(z)}{I_k(z_k)}\, X_{NDFT}[k], \tag{11.14}$$

where

$$I_k(z) = \prod_{\substack{i=0 \\ i \neq k}}^{N-1} \left(1 - z_i\, z^{-1}\right). \tag{11.15}$$

Write an M-function to compute the N-point NDFT of a sequence at a prescribed set of distinct N points z_k on the unit circle in the z-plane. The input to the function is the frequency points z_k and the input sequence, whereas, the output is the vector of NDFT samples. Write another M-function to evaluate the inverse NDFT using the Lagrange interpolation method. Using these functions evaluate the NDFT of sequences of various lengths and their inverse NDFTs.

An elegant application of the NDFT is in the efficient design of one-dimensional and two-dimensional FIR filters using a modified frequency-sampling approach, and in the design of beam-formers with prescribed nulls [Bag98].

Project 11.3   Warped Discrete Fourier Transform

As indicated in Eq. (11.1), the N-point DFT $X[k]$ of a length-N sequence $x[n]$ is given by the frequency samples of the z-transform $X(z)$ of $x[n]$ at N equally spaced points on the unit circle. The *warped discrete Fourier transform* (WDFT) can be employed to determine the N frequency samples of $X(z)$ at a warped frequency scale. The N-point WDFT $\check{X}[k]$ of $x[n]$ is given by the N equally spaced frequency samples on the unit circle of the modified z-transform $X(\hat{z})$ obtained by applying an allpass first-order spectral transformation to $X(z)$ [Mit98b]:

$$X(\hat{z}) = X(z)\bigg|_{z^{-1}=\frac{-\alpha+\hat{z}^{-1}}{1-\alpha\hat{z}^{-1}}}, \tag{11.16}$$

where $|\alpha| < 1$. Thus, the N-point WDFT $\check{X}[k]$ of $x[n]$ is given by

$$\check{X}[k] = X(\hat{z})\big|_{\hat{z}=e^{j2\pi k/N}}, \quad 0 \leq k \leq N-1. \tag{11.17}$$

Now, the modified z-transform $X(\hat{z})$ is given by

$$X(\hat{z}) = \sum_{n=0}^{N-1} x[n] \left(\frac{-\alpha + \hat{z}^{-1}}{1 - \alpha \hat{z}^{-1}} \right)^{-n} = \frac{P(\hat{z})}{D(\hat{z})}, \qquad (11.18)$$

where

$$P(\hat{z}) = \sum_{n=0}^{N-1} p[n]\, \hat{z}^{-n} = \sum_{n=0}^{N-1} x[n]\, (1 - \alpha \hat{z}^{-1})^{N-1-n} (-\alpha + \hat{z}^{-1})^n, \qquad (11.19)$$

and

$$D(\hat{z}) = \sum_{n=0}^{N-1} d[n]\, \hat{z}^{-n} = (1 - \alpha \hat{z}^{-1})^{N-1}. \qquad (11.20)$$

Note that the polynomial $D(\hat{z})$ depends only on the warping parameter α, whereas the polynomial $P(\hat{z})$ depends on both the input $x[n]$ and α. It follows from the above that the N-point WDFT $\check{X}[k]$ is simply given by $\check{X}[k] = P[k]/D[k]$ where $P[k]$ and $D[k]$ are, respectively, the N-point DFTs of the sequences $p[n]$ and $d[n]$ defined in Eqs. (11.19) and (11.20).

If we denote $\mathbf{P} = [p[0] \quad p[1] \quad \cdots \quad p[N-1]]^T$ and $\mathbf{X} = [x[0] \quad x[1] \quad \cdots \quad x[N-1]]^T$, then it can be shown that $\mathbf{P} = \mathbf{Q} \cdot \mathbf{X}$ where $\mathbf{Q} = [q_{r,s}]$ is a real $N \times N$ matrix whose first row is given by $q_{0,s} = \alpha^s, 0 \leq s \leq N - 1$, and whose first column is given by $q_{r,0} = {}^{N-1}C_r \alpha^r$. The remaining elements $q_{r,s}$ can be derived using the recursion relation

$$q_{r,s} = q_{r-1,s-1} + \alpha\, q_{r,s-1} - \alpha\, q_{r-1,s}. \qquad (11.21)$$

There are several interesting applications of the WDFT. For example, the warping parameter can be chosen to increase the frequency resolution at a selected portion of the angular frequency axis without changing the length N of the sequence and, if necessary, to determine a frequency sample at a specified frequency point. This is particularly attractive in spectral analysis of signals containing closely spaced sinusoids as here a short-length WDFT with an appropriate warping parameter can be used to provide the necessary spectral resolution. Another application of the WDFT is in designing FIR filters with tunable magnitude response.

Write a MATLAB function wdft to compute the WDFT of a sequence. The input data to your function are the input sequence vector and the warping parameter α, and the output is the vector of the WDFT samples. Using this function for $\alpha = -0.45$ perform a spectral analysis of a signal composed of two sinusoids located at 0.35 and 0.65 radians for different values of the DFT size. What is the smallest size N_1 of the WDFT necessary to resolve the two sinusoids? Now, perform a spectral analysis of the same signal using the function fft. What is the smallest size N_2 of the DFT necessary to resolve the two sinusoids? Compare the computational complexities of the WDFT- and the DFT-based approaches.

As a second application, write a MATLAB program to design and plot the magnitude response of a lowpass FIR filter with a specified passband edge and that of its transformed filter with a different passband edge. Use the function `kaiord` (see Section 7.5) to estimate the length N of the FIR filter and use `remez` to design the prototype FIR filter. Compute the WDFT of the impulse response of the prototype FIR filter for the given warping parameter. An inverse DFT of these frequency samples then yields the impulse response samples of the transformed filter. The input data to your program are the passband and stopband edges of the prototype, its length, and the warping parameter. Assume equal weights in the specified bands. Run your program for various values of the input data. Comment on your results.

11.3   FIR Filter Design and Implementation

In many applications FIR filters are preferred over IIR filters as they are guaranteed stable and do not exhibit limit cycles. However, the length of the FIR filter meeting a specified magnitude response is inversely proportional to the width of the transition band, and hence, in the case of a sharp cutoff filter, the length is often prohibitively large requiring an excessive number of multiplication and addition operations. In this section you will investigate the design of certain specific types of computationally efficient FIR filters.

Project 11.4   Interpolated FIR Filters

The *interpolated FIR filter* design approach is based on the realization of the filter as a cascade of two FIR filters with transfer functions $F(z^L)$ and $G(z)$, resulting in an overall transfer function given by

$$H(z) = F(z^L)\,G(z). \tag{11.22}$$

The filter $F(z^L)$ has a frequency response that is periodic with a period $2\pi/L$. The filter $G(z)$ attenuates the undesired passbands of this frequency response and keeps only the desired passband. In the time-domain, $F(z^L)$ has a sparse impulse response with $L-1$ zero-valued samples inserted between consecutive impulse response samples of $F(z)$, called the *model filter*. The filter $G(z)$, called the *interpolation filter*, thus peforms an interpolation of the impulse response of $F(z^L)$ and fills in the zero-valued samples.

The IFIR approach can be used to design narrowband lowpass, highpass, and bandpass filters [Neu84]. We consider here first the design of a narrowband lowpass filter $H(z)$ with a passband edge at ω_p, a stopband edge at ω_s, a passband ripple of δ_p, and a stopband ripple of δ_s. To this end, first design a wideband lowpass filter $F(z)$ with a passband edge at $L\,\omega_p$, a stopband edge at $L\,\omega_s$, a passband ripple of δ_p, and a stopband ripple of δ_s. Note that the maximum value of L is given by $\lfloor \pi/\omega_s \rfloor$. In practice, L should be chosen slightly smaller than this value to provide some separation between the replicated passbands in $F(z^L)$ to make the design of $G(z)$ simpler. To attenuate the replica at $\omega = \pi$, one can choose the FIR section $G_0(z) = \frac{1}{2}(1 + z^{-1})$ and to attenuate the replicas centered at ω_o, the FIR section $G_2(z) = \frac{1}{K}(1 + 2\cos\omega_o\, z^{-1} + z^{-2})$, can be used where the value of K is chosen to give a maximum gain value of 0 dB. To provide more attenuation in the undesired passbands, multiple use of the above two sections can be used. Note that once L

and the form of $G(z)$ have been selected, the effect of the magnitude response $|G(e^{j\omega})|$ of $G(z)$ on the overall magnitude response of the cascade can be compensated by using the inverse of $|G(e^{j\omega})|$ to modify the magnitude response of $F(z)$ in the desired passband of $H(z)$.

Using the function `remez`, design a length-99 IFIR lowpass filter with a passband edge at $\omega_p = 0.0404\pi$, a stopband edge at $\omega_s = 0.0556\pi$, and equal weights in the passband and the stopband. Design the model filter $F(z)$ for the $L = 2$ case first with length 49 and predistort its passband to compensate for the effect of $G(z)$. Compare the magnitude responses and the computational complexities of the IFIR filter and the FIR filter designed using the function `remez`. Next, design the IFIR filter using $L = 3$ and 4. Compare their performances with the IFIR filter designed using $L = 2$. Using the function a2dT described in Chapter 9, compare the coefficient sensitivity properties of the IFIR filters with that of the FIR filter designed using the function `remez`.

By a simple modification, the design approach outlined above can be used to design an IFIR highpass filter and an IFIR bandpass filter. For example, if $H(z)$ is a model lowpass FIR filter, by attenuating the replica of $H(z^2)$ at $\omega = 0$ by an interpolator $G_1(z) = (1 - z^{-1})$, one can design a highpass filter. Using this modification, design a narrowband highpass FIR filter with a passband edge $\omega_p = 0.9596\pi$, a stopband edge at $\omega_s = 0.94454\pi$, and equal weights in the passband and the stopband.

It is also possible to design a wideband FIR filter by designing a narrow-band delay complementary IFIR filter. However, in this case, the wideband filter must be a Type 1 linear-phase filter. For further details on the IFIR method see [Neu84].

The IFIR approach has been extended for the design of computationally efficient FIR bandpass filters [Neu87]. The method is based on modulating the impulse response $h[n]$ of the model lowpass filter $H(z)$ to generate a complex coefficient bandpass FIR filter $H_1(z)$ with passband centered at ω_o and a complex coefficient bandpass FIR filter $H_2(z)$ with passband centered at $-\omega_o$. The unwanted passbands of $H_1(z^L)$ and $H_2(z^L)$ are then attenuated with complex coefficient FIR interpolators $G_1(z)$ and $G_2(z)$, respectively. For a real coefficient $H(z)$, the coefficients of $H_2(z^L)$ are complex conjugate of those of $H_1(z^L)$. Likewise, the coefficients of $G_2(z)$ can be made to be complex conjugate of those of $G_1(z)$. Let $H(z) = H_{re}(z) + j\,H_{im}(z)$, and $G_1(z) = G_{re}(z) + j\,G_{im}(z)$, where $H_{re}(z)$, $H_{im}(z)$, $G_{re}(z)$, and $G_{im}(z)$, have real coefficients. Then the transfer function of the desired FIR bandpass filter is given by $\frac{1}{2}\{H_{re}(z)\,G_{re}(z) + H_{im}(z)\,G_{im}(z)\}$, resulting in a parallel structure.

Using the approach outlined above, write a MATLAB program to design a symmetric FIR bandpass filter with a center frequency at 0.87π, a passband width of 0.02π, and a transition bandwidth of 0.045π. The passband ripple is less than or equal to ± 0.025 dB and the minimum stopband attenuation is 51 dB [Neu87]. Plot the magnitude responses of this filter and compare them with those of an FIR bandpass filter designed using the function `remez`. Compare the hardware complexities of both filters.

Project 11.5   Frequency-Response Masking Approach

The frequency-response masking approach can be used to design linear-phase FIR filters with sharp transition bands [Lim86]. The basic idea behind this approach is as follows: Let $F_a(z)$ be a Type 1 lowpass FIR filter of order N with a frequency response

$$F_a(e^{j\omega}) = e^{-n\omega/2} R(\omega), \tag{11.23}$$

where $R(\omega)$ is its amplitude response. Let $\omega_{a,p}$ and $\omega_{a,s}$ denote, respectively, the passband and stopband edges of $F_a(z)$. The delay-complementaty filter $F_c(z)$ of $F_a(z)$ is given by

$$F_c(z) = z^{-N/2} - F_a(z), \tag{11.24}$$

which has an amplitude response $[1 - R(\omega)]$. Now consider the structure of Figure 11.1(a) where the filter $F_a(z^M)$ is a multiband filter with M passbands. Likewise, the filter $F_c(z^M)$ is also a multiband filter with an amplitude response that is complementary to that of $F_a(z^M)$. Any one or more of the passbands of $F_a(z^M)$ can be masked by using a masking filter $G_a(z)$, and similarly, one or more of the passbands of $F_c(z^M)$ can be masked by using a masking filter $G_c(z)$. By adding the outputs of the cascades $F_a(z^M)G_a(z)$ and $F_c(z^M)G_c(z)$ as shown in Figure 11.1(b), where $G_a(z)$ and $G_c(z)$ have been chosen appropriately, a wideband linear-phase FIR filter with a sharp transition band can be designed. For example, if the passband and stopband edges of $G_a(z)$ are at $(2m\pi+\omega_{a,p})/M$ and $[2(m+1)\pi - \omega_{a,s}]/M$, and if the passband and stopband edges of $G_c(z)$ are at $(2m\pi - \omega_{a,p})/M$ and $(2m\pi + \omega_{a,s})/M$, the structure of Figure 11.1(b) realizes a wideband lowpass filter with passband and stopband edges at

$$\omega_p = \frac{2m\pi + \omega_{a,p}}{M}, \qquad \omega_s = \frac{2m\pi + \omega_{a,s}}{M}, \tag{11.25}$$

where m is an integer less than M. Note that it is tacitly assumed here that the group delays of $G_a(z)$ and $G_c(z)$ are equal. If they are not equal, they should be made equal by adding an appropriate amount of leading delays to the one with the smaller group delay. Moreover, to avoid half sample delay, NM must be even. For further details on this method see [Lim86].

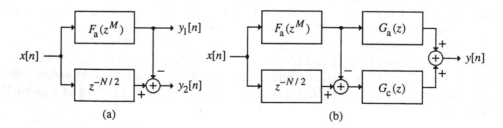

(a)   (b)

Figure 11.1 (a) Delay-complementary pair and (b) FIR filter structure employing frequency-masking approach.

Using MATLAB, design a linear-phase FIR filter based on the frequency-masking approach with the following specifications: passband edge at 0.3π, stopband edge at 0.305π, passband ripple of ± 0.1 dB and minimum stopband attenuation of 40 dB.

Project 11.6   FIR Filters with 0, +1, and −1 Coefficients

To simplify the hardware implementation, the design of FIR filters with 0, +1, and −1 coefficients have been proposed. In one such design approach, the overall filter structure is implemented as indicated in Figure 11.2 [Bat80]. The basic idea behind this approach is to convert the coefficients of a conventionally designed FIR filter $H(z)$ into a sequence with sample values 0, +1, and −1 by a delta modulation-like scheme. To this end, the impulse response $\{h[n]\}$ of $H(z)$ is up-sampled by a factor L, resulting in a transfer function $H_L(z) = H(z^L)$ with a corresponding impulse response $\{h_L[n]\}$. The objective is then to determine the sequence $\{w[n]\}$, with sample values 0, +1, and −1, whose running sum is a good approximation to the sequence $\{h_L[n]\}$ in the mean-square error sense, that is, for which the error

$$\mathcal{E}_L = \sum_{n=0}^{\infty} \left(h_L[n] - \Delta_L \sum_{\ell=0}^{n} w[\ell] \right)^2, \tag{11.26}$$

is minimized within a prescribed tolerance. In Eq. (11.26), Δ_L is a scaling factor corresponding to the stepsize in delta modulation. Further details on this method along with the optimization algorithm can be found in [Bat80].

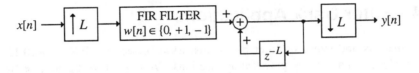

Figure 11.2   FIR filter implementation with 0, +1, and −1 coefficients.

Using MATLAB design a lowpass filter with a passband edge at 0.007π and a stopband edge at 0.023π based on the above approach. Use an up-sampling factor of 6.

Project 11.7   Running FIR Filter Structure

Various structures for the computationally efficient implementation of FIR filters using multirate techniques have been advanced by many authors. One such structure is shown in Figure 11.3 [Vet88]. Analyze this structure and show that it is alias-free with an overall transfer function

$$T(z) = z^{-1} \left[H_0(z^2) + z^{-1} H_1(z^2) \right]. \tag{11.27}$$

For a given FIR transfer function $T(z)$, one can obtain the two transfer functions $H_0(z^2)$ and $H_1(z^2)$ as follows:

$$\begin{aligned} H_0(z^2) &= \frac{1}{2}[T(z) + T(-z)], \\ H_1(z^2) &= \frac{1}{2} z [T(z) - T(-z)], \end{aligned} \tag{11.28}$$

If the length of $T(z)$ is $2K$, then the two filters $H_0(z)$ and $H_1(z)$ are of length K each.

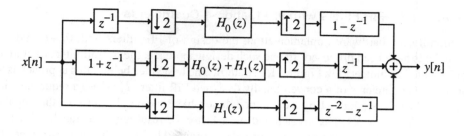

Figure 11.3 A fast running convolution structure.

Write a MATLAB program to determine from a given FIR transfer function $T(z)$ the transfer functions $H_0(z)$ and $H_1(z)$, and then simulate the structure of Figure 11.3. Verify the correctness of your simulation using the structure verification method given in R8.1. Use the function `strucver` described in Project 8.1.

11.4 Filter Bank Applications

Digital analysis and synthesis filter banks, discussed in sections R10.11-R10.13 in the previous chapter, have many practical applications. You will investigate here two such applications.

Project 11.8 Transmultiplexer

The *transmultiplexer* is an L-input, L-output, multirate structure formed by a synthesis filter bank at the input end and an analysis filter bank at the output end as shown in Figure 11.4. The transmultiplexer is designed to ensure that the kth output $y_k[n]$ is a reasonable replica of the kth input $x_k[n]$ for all values of k. There is *cross talk* between the kth and rth channels with $r \neq n$, if $x_k[n]$ contains contributions from $x_r[n]$. In a *perfect reconstruction* transmultiplexer, $y_k[n] = \alpha_k x_k[n - D]$, where α_k is a constant and D is a positive integer. The transmultiplexer is used in the *time-division multiplex* (TDM) to *frequency-division multiplex* (FDM) format conversion. Additional details on the theory, design, and application of the transmultiplexer can be found in [Mit98a].

In this project you will investigate the operation of a two-channel transmultiplexer with the following synthesis and analysis filters [Mit98a]:

$$G_0(z) = z^{-1} + z^{-2}, \qquad G_1(z) = z^{-1} - z^{-2},$$

$$H_0(z) = 1 + z^{-1}, \qquad H_1(z) = 1 - z^{-1}.$$

Write a MATLAB program to simulate this structure and process two arbitrary input sequences of length-20 each. Show that $y_k[n] = 2 x_k[n - 2]$ for $k = 0, 1$.

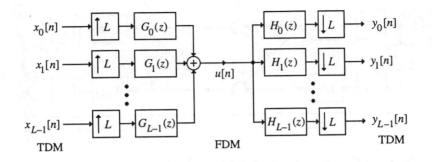

Figure 11.4 An L-channel transmultiplexer.

Project 11.9 Quadrature-Mirror Filter Bank

The L-channel *quadrature-mirror filter* (QMF) bank shown in Figure 11.5 consists of an L-channel analysis filter bank at the input end followed by an L-channel synthesis bank at the output end. The down-sampled output signals $u_k[n]$, called *subband signals*, of the analysis filters operate at a lower rate than the input to the QMF bank and can thus be processed more efficiently. The analysis and synthesis filters can be designed so that the QMF bank is alias-free and the output $y[n]$ is some type of replica of the input $x[n]$. A common application of the QMF bank is in the subband coding of speech, audio, image, and video signals.

In this project you will investigate the design and operation of a two-channel QMF bank shown in Figure 11.6. Analyze the structure and show that the analysis and the synthesis filters here are given by

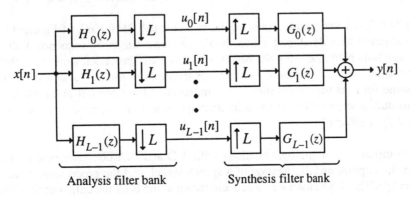

Analysis filter bank Synthesis filter bank

Figure 11.5 An L-channel quadrature-mirror filter bank.

$$H_0(z) = \frac{1}{2}\{\mathcal{A}_0(z^2) + z^{-1}\mathcal{A}_1(z^2)\}, \quad H_1(z) = \frac{1}{2}\{\mathcal{A}_0(z^2) - z^{-1}\mathcal{A}_1(z^2)\}, \quad \text{(11.29)}$$

$$G_0(z) = \frac{1}{2}\{\mathcal{A}_0(z^2) + z^{-1}\mathcal{A}_1(z^2)\}, \quad G_1(z) = \frac{1}{2}\{z^{-1}\mathcal{A}_1(z^2) - \mathcal{A}_0(z^2)\}. \quad \text{(11.30)}$$

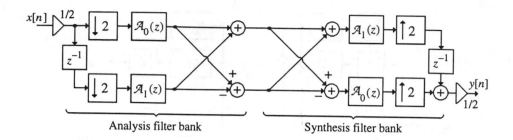

Figure 11.6 A magnitude-preserving two-channel quadrature-mirror filter bank.

Next show that the input-output relation of this QMF bank in the z-domain is given by

$$Y(z) = \frac{1}{4}\mathcal{A}_0(z^2)\mathcal{A}_1(z^2),\qquad(11.31)$$

indicating that it is an alias-free magnitude-preserving QMF bank.

Consider the fifth-order Butterworth half-band lowpass transfer function

$$H_0(z) = \frac{0.0527864045\,(1 + z^{-2})^5}{1 + 0.633436854\,z^{-1} + 0.05572809\,z^{-4}}.\qquad(11.32)$$

Show that it can be decomposed into the form

$$H_0(z) = \frac{1}{2}\left[\left(\frac{0.10557281 + z^{-2}}{1 + 0.10557281\,z^{-2}}\right) + z^{-1}\left(\frac{0.527864045 + z^{-2}}{1 + 0.527864045\,z^{-2}}\right)\right].\qquad(11.33)$$

From the above decomposition, determine the two allpass filters $\mathcal{A}_0(z)$ and $\mathcal{A}_1(z)$.

Write a MATLAB program to simulate the two-channel QMF bank of Figure 11.6 with the two allpass filters developed above. Next generate a length-50 sequence with a triangular magnitude response using the M-function `fir2` and process it using the QMF bank you have simulated. Determine the magnitude spectrum of the output and show that it is of the same form as that of the input. The initial conditions present in the allpass filters may introduce some visible distortion in the output magnitude spectrum. How would you minimize their effects?

If your computer has a microphone along with A/D and D/A converter boards, you may consider the capture and digitization of a speech signal, and the processing of this signal using the QMF bank you have simulated and then playing back the output of the filter bank to find out its quality.

11.5 Modulation and Demodulation

For the transmission of a low-frequency signal over a channel, the signal is transformed into a high-frequency signal by a modulation operation. The modulated high-frequency signal is demodulated at the receiving end and the desired low-frequency signal is then extracted by further processing. You will investigate here the operation of two types of modulation schemes: amplitude modulation and quadrature amplitude modulation.

Project 11.10   Amplitude Modulation

In amplitude modulation, the amplitude of a high-frequency sinusoidal signal $A \cos(\omega_o n)$, called the *carrier signal*, is varied by the low-frequency signal $x[n]$, called the *modulating signal*, generating a high-frequency signal, called the *modulated signal $y[n]$*, according to

$$y[n] = A\, x[n]\, \cos(\omega_o n). \tag{11.34}$$

The spectrum $X(e^{j\omega})$ of the modulating signal $x[n]$ is assumed to be band-limited to ω_m. Thus, amplitude modulation can be implemented by forming the product of the modulating signal with the carrier signal. The spectrum $Y(e^{j\omega})$ of $y[n]$ is given by

$$Y(e^{j\omega}) = \frac{A}{2}X(e^{j(\omega-\omega_o)}) + \frac{A}{2}X(e^{j(\omega+\omega_o)}). \tag{11.35}$$

The recovery of $x[n]$ from $y[n]$, called *demodulation*, assuming $\omega_o > \omega_m$, is carried out in two steps. First, the product of $y[n]$ with a sinusoidal signal of the same frequency as the carrier is formed. This results in

$$r[n] = y[n]\, \cos(\omega_o n) = A\, x[n]\, \cos^2(\omega_o n) = \frac{A}{2}x[n] + \frac{A}{2}x[n]\, \cos(2\omega_o n). \tag{11.36}$$

The product signal is therefore composed of the original modulating signal $x[n]$ scaled by a factor $\frac{1}{2}$ and an amplitude modulated signal with a carrier frequency of $2\omega_o$. The original modulating signal can now be recovered from $r[n]$ by passing it through a lowpass filter with a cutoff frequency ω_c, satisfying the relation $\omega_m < \omega_c < 2\omega_o - \omega_m$. The output of the filter is then a scaled replica of the modulating signal.

Figure 11.7 shows the block diagram representations of the amplitude modulation and demodulation schemes.

Write a MATLAB program to demonstrate the operation of the amplitude modulation and demodulation schemes.

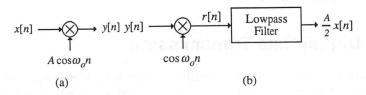

Figure 11.7 Schematic representations of the amplitude modulation and demodulation schemes: (a) modulator and (b) demodulator.

Project 11.11   Quadrature Amplitude Modulation

To understand the basic idea behind the *quadrature amplitude modulation* (QAM) approach, consider two band-limited low-frequency signals $x_1[n]$ and $x_2[n]$, with a bandwidth of ω_m each. The two modulating signals are individually modulated by the two carrier signals $A\cos(\omega_o n)$ and $A\sin(\omega_o n)$, respectively, and summed, resulting in a composite signal $y[n]$ given by

$$y[n] = A\,x_1[n]\,\cos(\omega_o n) + A\,x_2[n]\,\sin(\omega_o n). \tag{11.37}$$

Note that the two carrier signals have the same carrier frequency ω_o but have a phase difference of 90°. In general, the carrier $A\cos(\omega_o n)$ is called the *in-phase component* and the carrier $A\sin(\omega_o n)$ is called the *quadrature component*. The spectrum $Y(e^{j\omega})$ of the composite signal $y[n]$ is now given by

$$Y(e^{j\omega}) = \frac{A}{2}\left\{X_1(e^{j(\omega-\omega_o)}) + X_1(e^{j(\omega+\omega_o)})\right\}$$
$$+ \frac{A}{2j}\left\{X_2(e^{j(\omega-\omega_o)}) - X_2(e^{j(\omega+\omega_o)})\right\}. \tag{11.38}$$

To recover the original modulating signals, the composite signal is multiplied by both the in-phase and the quadrature components of the carrier separately resulting in two signals:

$$r_1[n] = y[n]\,\cos(\omega_o n), \qquad r_2[n] = y[n]\,\sin(\omega_o n). \tag{11.39}$$

Substituting $y[n]$ from Eq. (11.37) into Eq. (11.39), we obtain, after some algebra,

$$r_1[n] = \frac{A}{2}x_1[n] + \frac{A}{2}x_1[n]\cos(2\omega_o n) + \frac{A}{2}x_2[n]\sin(2\omega_o n),$$
$$r_2[n] = \frac{A}{2}x_2[n] + \frac{A}{2}x_1[n]\sin(2\omega_o n) - \frac{A}{2}x_2[n]\cos(2\omega_o n). \tag{11.40}$$

Lowpass filtering of $r_1[n]$ and $r_2[n]$ by filters with a cutoff at ω_m yields the two modulating signals. Figure 11.8 shows the block diagram representations of the quadrature amplitude modulation and demodulation schemes.

Write a MATLAB program to demonstrate the operation of the quadrature amplitude modulation and demodulation schemes.

11.6   Digital Data Transmission

Binary data are normally transmitted serially as a pulse train. However, in order to extract faithfully the information transmitted, the receiver requires complex equalization procedures to compensate for channel imperfection and to make full use of the channel bandwidth. To alleviate the problems encountered with the transmission of data as a pulse train, a multi-carrier modulation/demodulation scheme for digital data transmission is preferred.

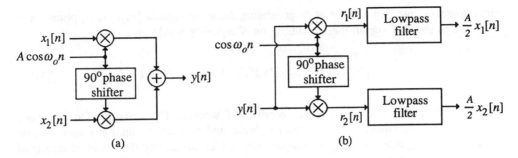

Figure 11.8 Schematic representations of the quadrature amplitude modulation and demodulation schemes: (a) modulator and (b) demodulator.

Project 11.12   Discrete Multitone Transmission

A widely used form of the multicarrier modulation is the *discrete multitone transmission* (DMT) scheme in which the modulation and demodulation processes are implemented via the discrete Fourier transform (DFT), efficiently realized using fast Fourier transform (FFT) methods [Cio91], [Pel80]. To understand the basic idea behind the DMT scheme, consider the transmission of two $M - 1$ real-valued data sequences, $\{a_k[n]\}$ and $\{b_k[n]\}$, $1 \leq k \leq M - 1$ operating at a sampling rate of F_T. Define a new set of complex sequences $\alpha_k[n]$ of length $N = 2M$ according to

$$\alpha_k[n] = \begin{cases} 0, & k = 0, \\ a_k[n] + j\, b_k[n], & 1 \leq k \leq \frac{N}{2} - 1, \\ 0, & k = \frac{N}{2}, \\ a_{N-k}[n] + j\, b_{N-k}[n], & \frac{N}{2} + 1 \leq k \leq N - 1. \end{cases} \qquad (11.41)$$

By applying an inverse DFT, the above set of N sequences is transformed into another new set of N signals $\{u_\ell[n]\}$, given by

$$u_\ell[n] = \frac{1}{N} \sum_{k=0}^{N-1} \alpha_k[n]\, W_N^{-\ell k}, \qquad 0 \leq \ell \leq N - 1, \qquad (11.42)$$

where $W_N = e^{-j2\pi/N}$. Note that the method of generation of the complex sequence $\{\alpha_k[n]\}$ ensures that its IDFT $\{u_\ell[n]\}$ will be a real sequence. Each of these N signals is then up-sampled by a factor of N and time-interleaved, generating a composite signal $\{x[n]\}$ operating at a rate $N\,F_T$, which is assumed to be equal to $2\,F_c$. The composite signal is converted into an analog signal $x_a(t)$ by passing it through a D/A converter followed by an analog reconstruction filter. The analog signal $x_a(t)$ is then transmitted over the channel.

At the receiver, the received analog signal $y_a(t)$ is passed through an analog anti-aliasing filter and then converted into a digital signal $\{y[n]\}$ by a sample-and-hold (S/H) circuit followed by an A/D converter operating at a rate $N\,F_T = 2\,F_c$. The received digital signal is then de-interleaved by a delay chain containing $N - 1$ unit delays whose outputs are next down-sampled by a factor of N generating the set of signals $\{v_\ell[n]\}$. Application of the DFT to these N signals then results in the N signals $\{\beta_k[n]\}$ given by

$$\beta_k[n] = \sum_{\ell=0}^{N-1} \alpha_k[n] W_N^{\ell k}, \quad 0 \le k \le N-1. \tag{11.43}$$

Figure 11.9 shows schematically the overall DMT scheme. If we assume the frequency response of the channel to have a flat passband, and assume the analog reconstruction and anti-aliasing filters to be ideal lowpass filters, then neglecting the nonideal effects of the A/D and the D/A converters, we can assume $y[n] = x[n]$. In this case, the interleaving circuit of the DMT structure at the transmitting end connected to the de-interleaving circuit at the receiving end is an ideal transmultiplexer structure and, hence, it follows that

$$\begin{aligned} v_k[n] &= u_{k-1}[n], \quad 1 \le k \le N-2, \\ v_0[n] &= u_{N-1}[n], \end{aligned} \tag{11.44}$$

or, in other words,

$$\begin{aligned} \beta_k[n] &= \alpha_{k-1}[n-1], \quad 1 \le k \le N-2, \\ \beta_0[n] &= \alpha_{N-1}[n]. \end{aligned} \tag{11.45}$$

Write a MATLAB program to demonstrate the operation of the DMT data scheme.

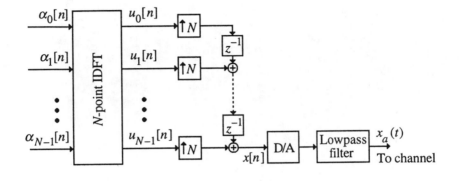

(a)

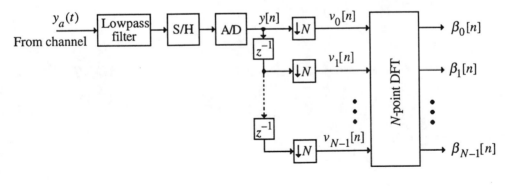

(b)

Figure 11.9   The DMT scheme. (a) transmitter and (b) receiver.

Introduction to MATLAB A

MATLAB is a powerful high-level programming language for scientific computations. It is very easy to learn and use in solving numerically complex engineering problems. The exercises in this book have been written assuming you are not proficient in MATLAB. However, some basic concepts of MATLAB are included here for a quick review to facilitate your understanding of the programs and for performing the exercises. A more detailed review can be found in [Mat94], [Mat96], [Mat97].

MATLAB consists of functions that are either built into the interpretor or available as M-files, with each containing a sequence of program statements that execute a certain algorithm. A complete new algorithm can be written as a program containing only a few of these functions and can be saved as another M-file.

MATLAB works with three types of windows on your computer screen. These are the *Command window*, the *Figure window* and the *Editor window*. The Command window has the heading Command, the Figure window has the heading Figure No. 1, and the Editor window has the heading showing the name of an opened existing M-file or Untitled if it is a new M-file under construction. The Command window also shows the prompt >> indicating it is ready to execute MATLAB commands. Results of most printing commands are displayed in the Command window. This window can also be used to run small programs and saved M-files. All plots generated by the plotting commands appear in a Figure window. Either new M-files or old M-files are run from the Command window. Existing M-files can also be run from the Command window by typing the name of the file.

In the remaining part of this appendix we illustrate the use of some of the most commonly used functions and review some fundamental concepts associated with MATLAB.

A.1 Number and Data Representation

MATLAB uses conventional decimal notations to represent numbers with a leading minus sign for negative numbers. The approximate range of numbers that can be represented is from 10^{-308} to 10^{308}. Very large or very small numbers can be represented using exponents. Typical examples of valid number representations are as follows:

  1234.56789   123456.789E − 2   1.23456789e3   − 1234.56789

There should be no blank space before the exponent.

The data in MATLAB are represented in the form of a rectangular matrix that does not require dimensioning. Its elements can be either real or complex numbers. Thus, a one-dimensional discrete-time signal can be represented either as a row or a column vector. For example the row vector indexMATLAB! data representation in

$$x = [3.5 + 4*j \quad -2.1 - 7.4*j \quad 1.05 - 0.8*j \quad 0 \quad 9.2*j];$$

denotes a complex-valued signal x of length 5. Note the use of square brackets to indicate that x is a rectangular matrix. Note also that the imaginary part of a complex number is represented using the operator $*$ and the letter j. An alternate form of representation of the imaginary part uses the letter i instead of the letter j. The real and imaginary parts of a complex number should be entered without any blank spaces on either side of the $+$ or $-$ sign as indicated above. The elements in the row of a matrix can also be separated by commas as indicated below:

$$x = [3.5 + 4*j, \quad -2.1 - 7.4*j, \quad 1.05 - 0.8*j, \quad 0, \quad 9.2*j];$$

The semicolon ; at the end of the square brackets ensures that the data are not printed in the command window after they have been entered. If the above data were entered without the semicolon, MATLAB would print in the Command window

```
x =
   Columns 1 through 4
   3.5000 + 4.0000i    -2.1000 - 7.4000i    1.0500 - 0.8000i    0
   Column 5
   0 + 9.2000i
```

Alternately, if needed, the actual value of x can be printed by typing x in the Command window.

The elements of a matrix can be entered in two different ways. The rows can be typed on a single line separated with semicolons or on different lines. For example, the 3×4 matrix A

$$A = \begin{bmatrix} 1 & 3 & 5 & 7 \\ 2 & 4 & 6 & 8 \\ 9 & 11 & 13 & 15 \end{bmatrix}$$

can be entered either as

$$A = [1 \quad 3 \quad 5 \quad 7; \quad 2 \quad 4 \quad 6 \quad 8; \quad 9 \quad 11 \quad 13 \quad 15];$$

or as

$$A = [1 \quad 3 \quad 5 \quad 7$$
$$2 \quad 4 \quad 6 \quad 8$$
$$9 \quad 11 \quad 13 \quad 15];$$

The indexing of vectors and matrices in MATLAB begins with 1. For example, x(1) in the above vector x is 3.5000 + 4.0000i, x(2) is -2.1000 - 7.4000i, and so on. Similarly, the first element in the first row of a matrix A is given by A(1,1), the second element in the first row is given by A(1,2), and so on. The index cannot be less than 1 or greater than the dimension of the vector or matrix under consideration.

The size of an array in the workspace of MATLAB can be determined by using the function size. For example, by typing size(x) we obtain the result

```
ans =
    1    5
```

The length of a vector can also be found by using the function length. For example, typing length(x) yields the result

```
ans =
    5
```

The array transpose operation is implemented using the operator .'. Thus the transpose of X is given by the expression X.'. If X is a matrix with complex-valued elements, X' is the complex conjugate transpose of X, whereas if X is a matrix with real-valued elements, X' is the transpose of X.

The data vectors and matrices in MATLAB can be labeled by a collection of characters including the numbers, such as x, x1, X, X1, XY, and so on. It should be noted that MAT-LAB normally differentiates between lowercase and uppercase letters.

Example A.1

Let X denote the 3×4 real-valued matrix entered by typing

$$X = [1 \quad 2 \quad 3 \quad 4; \quad 5 \quad 6 \quad 7 \quad 8; \quad 9 \quad 10 \quad 11 \quad 12];$$

Then typing X in the Command window results in the display of

```
ans =
    1     2     3     4
    5     6     7     8
    9    10    11    12
```

and typing X' we get

```
ans =
    1     5     9
    2     6    10
    3     7    11
    4     8    12
```

Consider next a 2×3 complex-valued matrix Y entered as

```
        Y = [1+2*i, 3-4*i, 5+6*i; 7-8*i, 9+10*i, 11-12*i];
```

Typing of Y yields

```
Y =
    1.0000 + 2.0000i    3.0000 - 4.0000i     5.0000 + 6.0000i
    7.0000 - 8.0000i    9.0000 + 10.0000i   11.0000 - 12.0000i
```

whereas typing Y′ we get

```
ans =
    1.0000 - 2.0000i     7.0000 + 8.0000i
    3.0000 + 4.0000i     9.0000 - 10.0000i
    5.0000 - 6.0000i    11.0000 + 12.0000i
```

To obtain the transpose of Y we type Y.′ resulting in

```
ans =
    1.0000 + 2.0000i     7.0000 - 8.0000i
    3.0000 - 4.0000i     9.0000 + 10.0000i
    5.0000 + 6.0000i    11.0000 - 12.0000i
```

A.2   Arithmetic Operations

Two different types of arithmetic operations are available in MATLAB for the manipulation of stored data, as indicated below where X and Y denote two different matrices. If X and Y are of the same dimensions, the *addition* of X and Y is implemented by the expression X + Y. The addition operation + can also be used to add a scalar to a matrix. Likewise, the *subtraction* of Y from X is implemented by the expression X − Y. The subtraction operation − can also be used to subtract a scalar from a matrix.

If the number of columns of X is the same as the number of rows of Y, the *matrix multiplication* X*Y yields the linear algebraic product of X and Y. The multiplication operation * can also be used to multiply a matrix by a scalar. If X and Y have the same dimensions, X.*Y is an *array multiplication* forming the element-by-element product of X and Y.

If Y is a square matrix and X is a matrix with the same number of columns as that of Y, then the *matrix right division* X/Y is equivalent to X*inv(Y) where inv(Y) denotes the inverse of Y. The right division operation X/Y can also be carried out if one of them is a scalar. If Y is a square matrix and X is a matrix with the same number of rows as that of Y, then the *matrix left division* Y\X is equivalent to inv(Y)*X. If X and Y are of the same dimension, the *array right division* is implemented using the expression X./Y , resulting in a matrix whose (r,s)-th element is given by X(r,s)/Y(r,s).

If multiple operations are employed in a statement, the usual precedence rules are followed in evaluating the expression. However, parentheses can be used to change the precedence of operations.

Arithmetic operations on matrices are illustrated in the following example.

Example A.2

Let X = [1 2 3; 4 5 6] and Y = [12 11 10; 9 8 7]. Then X+Y yields

ans =
        13   13   13
        13   13   13

and X-Y yields

ans =
      -11   -9   -7
       -5   -3   -1

The result of the operation X+3 is given by

ans =
        4   5   6
        7   8   9

whereas the result of the operation X*3 yields

ans =
        3   6   9
       12   15   18

The statement X.*Y develops the answer

ans =
       12   22   30
       36   40   42

Typing X*Y' we obtain the result

ans =
       64   46
      163   118

and typing X'*Y we arrive at

ans =

```
    48      43      38
    69      62      55
    90      81      72
```

Consider the two matrices X = [1 2 3; 4 5 6; 7 8 9] and Y = [1 1 2; 2 2 3; 1 3 4].

Then X/Y yields

```
ans =
     0.5000          0      0.5000
    -2.5000     3.0000      0.5000
    -5.5000     6.0000      0.5000
```

and Y\X results in

```
ans =
     0      0      0
     5      4      3
    -2     -1      0
```

A.3   Relational Operators

The relational operators in MATLAB <, <=, >, >=, ==, and ~=, represent the comparison operations *less than*, *less than or equal to* ($\leq$), *greater than*, *greater than or equal to* ($\geq$), *equal to*, and *not equal to* ($\neq$), respectively. Element-by-element comparisons between two matrices of the same size are carried out using these operators with the result appearing as a matrix of the same size whose elements are set to 1 when the relation is TRUE and set to 0 when the relation is FALSE. In the case of complex-valued matrices, the operators <, <=, >, and >= are applied to compare only the real parts of each element of the matrices, whereas the operators == and ~= are applied to compare both real and imaginary parts.

We illustrate the use of these operators in the following example.

Example A.3

Consider the two matrices C = [1 2 3; 4 5 6] and D = [1 7 2; 6 5 1]. Then the results of applying the above relational operators on C and D are indicated below:

```
C < D   =     0     1     0
              1     0     0

C > D   =     0     0     1
              0     0     1
```

```
C <= D  =   1   1   0
            1   1   0

C >= D  =   1   0   1
            0   1   1

C == D  =   1   0   0
            0   1   0

C ~= D  =   0   1   1
            1   0   1
```

A.4   Logical Operators

The three logical operators in MATLAB, &, |, and ~, perform the logical AND, OR, and NOT operations. When applied to matrices, they operate element-wise, with FALSE represented by a 0 and TRUE represented by a 1. We illustrate the use of these operators in the following example.

Example A.4

Consider the two matrices A = [1 1 0 1] and B = [0 1 0 0]. The results of applying the above logical operators on A and B are illustrated below:

```
A & B   =   0   1   0   0

A | B   =   1   1   0   1

  ~A    =   0   0   1   0
```

A.5   Control Flow

The control flow commands of MATLAB are break, else, elseif, end, error, for, if, return, and while. These commands permit the conditional execution of certain program statements. The command for is used to repeat a group of program statements a specific number of times. The command if is used to execute a group of program statements conditionally, and the command while can be used to repeat program statements an indefinite number of times. The statements following the commands for, while, and if must be terminated with the command end. The command break is used to terminate the execution of a loop. The commands else and elseif are used with the command if to provide conditional breaks inside a loop. The command error is employed to display error message and abort functions.

The use of these commands is illustrated in the following examples.

Example A.5

Consider the generation of a length-N sequence x of integers beginning with a specified first element x(1) and with each succeeding element increasing linearly by a specified positive integer D. The MATLAB program generating and displaying this sequence is given below:

```
N = 10;
D = 3;
x = [5 zeros(1,N-1)];
for k = 2:N
    x(k) = x(k-1) + D;
end
disp('The generated sequence is');disp(x)
```

Example A.6

Now consider the generation of a length-N sequence x of integers beginning with a specified first element x(1) and with each succeeding element increasing linearly by a specified positive integer D until an element is equal to R*D + x(1), where R is a positive integer, and then each succeeding element decreasing linearly by an amount D until an element is equal to x(1), and then repeating the process. A MATLAB program generating this sequence is as follows:

```
N = 15; D = 3;
x = [5 zeros(1,N-1)];
for k = 2:N
    x(k) = x(k-1) + D;
    if x(k) == 3*D + x(1)
        D = -abs(D);
    elseif x(k) == x(1)
        D = abs(D);
    end
end
disp('The generated sequence is');disp(x)
```

Example A.7

The following program illustrates the use of the command break. The program develops the sum of a series of numbers beginning with a specified initial value y, with each succeeding number increasing by a fixed positive increment D; stops the addition process when the total sum exceeds 100; and then displays the total sum.

```
y = 5; D = 3;
while 1
    y = y + D;
```

```
    if y > 100, break, end
end
disp('y is');disp(y)
```

A.6   Special Characters and Variables

MATLAB uses a number of special characters and words to denote certain items exclusively. These characters and words should not be used for any other purpose. For example, if the letter i or the letter j is used as a variable, it cannot be used to represent the imaginary part of a complex number. Hence, it is a good practice to restrict either the letter i or the letter j exclusively for the representation of the imaginary part of complex numbers.

There are several permanent variables that cannot be cleared by the user and should not be used to denote any other quantities. The word pi is used to denote π. Thus, sin(pi/4) yields 0.70710678118655, which is equal to $\sqrt{2}$. The variable eps is equal to 2^{-52} and is a tolerance for determining precision of certain computations such as the rank of a matrix. It can be set to any other value by the user. NaN represents *Not-a-Number*, which is obtained when computing mathematically undefined operations such as 0/0 and $\infty - \infty$. inf represents $+\infty$ and results from operations such as dividing by zero, for example, 2/0, or from overflow, for example, e^{1000}. The variable ans stores the results of the most recent operation.

The square brackets [] are used to enter matrices and vectors. The elements of a matrix can be separated by spaces or commas ,. A semicolon ; indicates the end of a row in a matrix. It is also used to suppress printing. The precedence in arithmetic expressions can be indicated using the parentheses (). The parentheses are also employed to enclose the indices of an array and arguments of functions. The operator notation for transpose of an array is '. However, two such symbols can be used to denote a quote. For example, 'dsp program' is a vector containing the ASCII codes of the characters inside the quotation marks. Any text following a percent symbol % denotes a comment and is not treated as a program statement.

The colon symbol : has many different applications in MATLAB. It is used to generate vectors, subscript matrices, and perform iterations of a block of commands. For example, x = M:N generates the vector

```
    x = [M    M+1    M+2    ...    N],
```

if M < N. However x = M:N is an empty matrix, denoted by [], if M > N. The command x = M:k:N generates the vector

```
    x = [M    M+k    M+2k    ...    N],
```

where k can be a positive or a negative integer. Note that x = M:k:N generates the empty matrix [] if k > 0 and M > N or if k < 0 and M < N.

The colon can also be employed to select specific rows, columns, and elements of a matrix or a vector. For example, Y(:,N) represents the Nth column of Y. Likewise, the Mth row of Y is represented by Y(M,:). Y(:,M:N) is equivalent to Y(:,M), Y(:,M+1), ... , Y(:,N). Finally, Y(:) is equivalent to a column vector formed by concatenating the columns of Y.

A.7   Output Data Format

All arithmetic operations in MATLAB are performed in double precision. However, different formats can be used to display the result of such operations in the Command window. If all results are exact integers, they are displayed as such without any decimal points. If one or more data elements are not integers, the results can be displayed with various precision. format short displays five significant decimal digits and is the default format. format short e displays five significant decimal digits with two positive or negative decimal exponents. format long shows results in 15 significant decimal digits, while format long e adds two positive or negative decimal exponents to 15 significant decimal digits. There are three other formats for displaying results. However, these are not that useful in signal processing applications.

A.8   Graphics

MATLAB includes high-level graphics capability for displaying the results of a computation. In most situations, we shall be concerned with two-dimensional (2-D) graphics and will use three-dimensional (3-D) graphics in some special cases. For 2-D graphics, plotting can be done in various forms with either linear or logarithmic scales for one or both axes of the plots. Grid lines can be added to the plots along with labels for the two axes and a title on top of the plot. Text can be placed anywhere on the graph using a mouse or specifying the starting position through commands in the program. Moreover, by including appropriate symbols in the argument of the plotting command, specific line styles, plot symbols, and colors can be displayed in the graph.

For 3-D data, plotting can also be done in various forms with either linear or logarithmic scales for one or two or all three axes of the plots. For example, lines and points can be plotted in three dimensions. Contour plots, 3-D perspective plots, surface plots, pseudocolor plots, and so forth can also be generated.

The M-file in the following section illustrates the use of several graphics commands.

A.9   M-Files: Scripts and Functions

An M-file is a sequence of MATLAB statements developed using a word processor or a text editor and saved with a name that must be in the form filename.m. The names of M-files must begin with a letter followed by at most 18 letters and/or digits (or underscores).

However certain characters, such as hyphen – and decimal point ., are not allowed in the names. Also, do not use the names of existing M-files. An M-file can include references to other existing M-files.

Each statement of a new program is typed in the Editor window line by line as ASCII text files and can be edited using the text editor or the word processor of your computer.[1] The complete program can then be saved as an M-file.

There are two types of M-files: *scripts* and *functions*. A function file must contain the word function in the first line of all program statements. Arguments in a function file may be passed from another M-file, and all variables used inside the function file are local.

The script file makes use of workspace data globally. The first line of a function file contains the word function and does not make use of workspace data globally. The variables defined inside a function file are manipulated locally, and the arguments of the file may be passed. When a function is completed, all local variables are lost. Only values specifically passed out are retained.

A simple example of a function file runsum is given below.

```
function y = runsum(x)
% Computes the mean of a vector x
L = length(x);
y = sum(x)/L;
```

A simple example of a script file lowpass.m follows.

```
% Script M-file lowpass.m
% Program to Perform Lowpass Filtering
% Using Three-Point Averaging of a Random Signal
% Program uses the function file runsum
z = zeros(1,11);data = randn(size(z));
u = [zeros(1,3) data];
N = 3; % N is the filter length
for k = 1:10;
    w = u(k:k+N);
    z(k) = runsum(w);
end
n = 0:10;
% Plot the noise in solid line and
% the smoothed version in dashed line
plot(n,data,'r-',n,z,'b--');grid
xlabel('Time index n');
```

---

[1]It should be noted that there is no built-in text editor on a UNIX workstation. Hence, the program must be written using other types of editors.

```
ylabel('Amplitude');
gtext('Noisy data');gtext('Smoothed data');
```

The plot generated by executing the M-file `lowpass.m` is shown in Figure A.1.

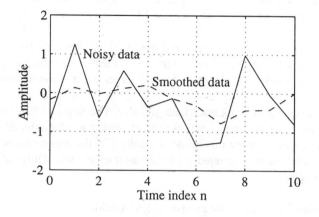

Figure A.1 Signal smoothing example.

Note that the function file `runsum` uses the built-in function `sum`. Likewise, the script file `lowpass.m` uses the function file `runsum`.

A.10 MAT-Files

Data generated by a MATLAB program can be saved as a binary file, called a MAT-file, for later use. For example, the noisy data generated by executing the program `lowpass.m` of the previous section can be saved using the command

```
save noise.mat data
```

Later it can be retrieved using the command

```
load noise
```

for use in another MATLAB session.

The data can also be saved in ASCII form. For example, after execution of the program `lowpass.m`, we can generate a 2×11 matrix containing the noisy data and the smoothed data using the command

```
result = [noise; z];
```

and then save the matrix `result` in ASCII form using the command

```
save tempo.dat result -ascii
```

The stored data can later be retrieved using the command

```
load tempo
```

A.11   Printing

To develop a hardcopy version of the current Figure window, the command `print` can be used. There are many versions of this command. See the *MATLAB Reference Guide* [Mat94] for details. In a PC or a Mac environment, a figure can also be copied to the clipboard and then inserted into a word processing document. This approach permits generating a smaller size figure and also pasting the figure on to a text.

A.12   Diagnostics and Help Facility

MATLAB has very good diagnostic capabilities, making it easier to correct any errors detected during execution. If any program statement has errors, the execution of the program will stop with a self-evident error message appearing in the Command window. For example, entering the real number 1.23456789e3 with a space before the exponent will result in the error message

```
???  1.23456789   e3
               |
Missing operator, comma, or semi-colon.
```

Entering the real number 1.23456789e3 with a colon in place of the decimal point as 1:23456789e3 will cause the error message

```
???  Error using ==> colon
Maximum variable size allowed by the program is exceeded.
```

MATLAB provides online information for most topics through the command `help`. If `help` is typed in the Command window with no arguments, a list of directories containing MATLAB files are displayed in the window. For help on specific M-files or directories, the name of the file or the directory should be used as an argument. For example typing `help runsum` results in

```
Computes the mean of a vector x
```

Likewise, typing `help lowpass` yields

```
A Script M-File to Perform Lowpass Filtering
Using Three-Point Averaging
Program uses the function file runsum
```

A list of variables in the workspace can be obtained by typing `who`. To obtain information about the size of the variables, use the command `whos`. Other useful commands are `what`, `which`, `lookfor`, `echo`, and `pause`.

The command `what` lists all files in the current directory, whereas the command `what dryname` lists the files in the directory named `dryname` on MATLAB's search path. The command `which` is used to locate functions and files on MATLAB's search path. The command `lookfor abc` searches through all help entries on MATLAB's search path and looks for the string `abc` in the first comment line. The command `echo` is useful for debugging a new program and is used to list all M-files being invoked during the execution of a program. There are several versions of this command. See the *MATLAB Reference Guide* [Mat94] to determine the appropriate ones for you to use. The command `pause` stops program execution temporarily at the point it is invoked; the execution can be restarted at that point by pressing any key on the keyboard. This command is particularly useful when the program is generating a large number of plots and each plot can be examined or copied individually if the command `pause` is inserted after each plotting command.

A.13 Remarks

Even though MATLAB uses double precision arithmetic, numerical approximations used in the computations may generate errors in the results. Care must be taken in such cases to interpret the results correctly. As an example, the computation of the expression $1 - 0.1 - 0.3 - 0.2 - 0.2 - 0.1 - 0.1$ yields $5.551115123125783e-17$ in the output `format long` when the result should have been ideally equal to 0. On the other hand, a slight change in the expression to $1 - (0.1 + 0.3 + 0.2 + 0.2 + 0.1 + 0.1)$ yields the correct result 0.

A Summary of MATLAB Commands Used

We provide below a quick review with a brief description of all MATLAB functions used in this book. For additional details on these functions, use the help command.

Function	Description
abs	Computes the absolute value
angle	Computes the phase angle in radians
axis	Sets manual scaling of axes on plots
blackman	Generates the Blackman window coefficients
break	Terminates the execution of loops
butter	Designs digital and analog Butterworth filters of all four types
buttord	Selects the minimum order of the digital or analog Butterworth transfer function
ceil	Rounds to the nearest integer towards $+\infty$
cheb1ord	Selects the minimum order of the digital or analog Type 1 Chebyshev transfer function
cheb2ord	Selects the minimum order of the digital or analog Type 2 Chebyshev transfer function
chebwin	Generates the Dolph-Chebyshev window coefficients
cheby1	Designs digital and analog Type 1 Chebyshev filters of all four types
cheby2	Designs digital and analog Type 2 Chebyshev filters of all four types
clf	Deletes all objects from the current figure
conj	Computes the complex conjugate
conv	Performs the multiplication of two polynomials
cos	Computes the cosine
decimate	Decreases the sampling rate of a sequence by an integer factor
deconv	Performs polynomial division
disp	Displays text or a matrix on the screen
echo	Echoes M-files during execution
ellip	Designs digital and analog elliptic filters of all four types
ellipord	Selects the minimum order of the digital or analog elliptic transfer function
else	Delineates an alternate block of statements inside an if loop
elseif	Conditionally executes a block of statements inside an if loop

end	Terminates a loop
eps	Indicates floating-point relative accuracy
error	Displays an error message
exp	Computes the exponential
fft	Computes the discrete Fourier transform coefficients
filter	Filters data with an IIR ir FIR filter implemented in the transposed direct form II structure
filtfilt	Performs zero-phase filtering of data
fir1	Designs linear-phase FIR filters of all four types using the windowed Fourier series method
fir2	Designs linear-phase FIR filters with arbitrary magnitude responses using the windowed Fourier series method
fix	Rounds towards zero
fliplr	Flips matrices left to right
flops	Computes the cumulative number of floating-point operations
for	Used for repeated execution of a block of statements a specific number of times
format	Controls the format of the output display
freqs	Computes the complex frequency response of an analog transfer function at specified frequency points
freqz	Computes the complex frequency response of a digital transfer function at specified frequency points
function	Used to generate new M-functions
grid	Adds or deletes grid lines to or from the current plot
grpdelay	Computes the group delay of a digital transfer function at specified frequency points
gtext	Places a text on a graph with the aid of a mouse
hamming	Generates the Hamming window coefficients
hanning	Generates the von Hann window coefficients
help	Provides online documentation for MATLAB functions and M-files
hold	Holds the current graph
if	Conditionally executes statements
ifft	Computes the inverse discrete Fourier transform coefficients
imag	Determines the imaginary part of a complex number or matrix
impz	Computes a specific number of the impulse response coefficients of a digital transfer function
input	Requests data to be supplied by the user
interp	Increases the sampling rate of a sequence by an integer factor
inv	Determines the inverse of a matrix
kaiser	Determines the Kaiser window coefficients
kaiserord	Detemines the filter order and the parameter β of a Kaiser window
latc2tf	Determines the transfer function from the specified lattice parameters and the feed-forward coefficients of the Gray-Markel realization

legend	Inserts a legend on the current plot using the specified strings as labels
length	Determines the length of a vector
linspace	Generates linearly spaced vectors
load	Retrieves saved data from the disk
log10	Computes the common logarithm
lookfor	Provides keyword search through all help entries
max	Detemines the largest element of a vector
min	Detemines the smallest element of a vector
NaN	Not-a-number
nargin	Indicates the number of arguments inside the body of a function M-file
num2str	Converts a number to its string representation
ones	Generates a vector or a matrix of with element value 1
pause	Halts execution temporarily until user presses any key
pi	Returns the floating-point number nearest to π
plot	Generates linear2-D plots
poly2rc	Determines the coefficients in the cascade realization of an IIR allpass transfer function
rand	Generates random numbers and matrices uniformly distributed in the interval (0,1)
randn	Generates random numbers and matrices normally distributed with zero mean and unity variance
real	Determines the real part of a complex number or matrix
rem	Determines the remainder of a matrix divided by another matrix of same size
remez	Designs linear-phase FIR filters using the Parks-McClellan algorithm
remezord	Determines the approximate order, normalized band edges, frequency band magnitude levels, and weights to use with the remez command
resample	Changes the sampling rate of a sequence by a rational number
residue	Determines the partial-fraction of a discrete-time transfer function expressed as a ratio of polynomials in z
residuez	Determines the partial-fraction of a discrete-time transfer function expressed as a ratio of polynomials in z^{-1}
return	Causes a return to the keyboard or to the invoking function
roots	Determines the roots of a polynomial
save	Saves workspace variables on a disk
sawtooth	Generates a sawtooth wave with a period 2π
sign	Implements the signum function
sin	Determines the sine
sinc	Computes the sinc function of a vector or array
size	Returns the matrix dimensions
sqrt	Computes the square root
square	Generates a square wave with a period 2π

stairs	Draws a stairstep graph
stem	Plots the data sequence as stems from the x axis terminated with circles for the data value
subplot	Breaks figure window into multiple rectangular panes for the display of multiple plots
sum	Determines the sum of all elements in a vector
tf2latc	Determines the lattice-parameters and the feed-forward coefficients in the Gray-Markel realization of an IIR transfer function
tf2zp	Determines the zeros, poles, and gains of the specified transfer function
title	Write specified text on the top of the current plot
unwrap	Eliminates jumps in phase angles to provide smooth transition across branch cuts
what	Provides directory listing of files
which	Locates functions and files
while	Repeats statements an indefinite number of times
who	Lists the current variables in the memory
whos	Lists the current variables in the memory, their sizes, and whether they have non-zero imaginary parts
xlabel	Write specified text below the x-axis of the current 2-D plot
ylabel	Write specified text below the y-axis of the current 2-D plot
zeros	Generates a vector or a matrix with element 0
zp2sos	Determines an equivalent second-order representation from a specified zero-pole-gain representation
zp2tf	Determines the numerator and the denominator coefficients of a transfer function from its specified zeros, poles, and gains
zplane	Displays poles and zeros in the z-plane

References

[Abr72] M. Abramowitz and I.A. Segun, editors. *Handbook of Mathematical Functions*. Dover Publications, New York NY, 1972.

[Bag98] S. Bagchi and S.K. Mitra. *Nonuniform Discrete Fourier Transform and Its Signal Processing Applications*. Kluwer, Boston MA, 1998.

[Bat80] M.R. Bateman and B. Liu. An approach to programmable CTD filters using coefficients 0, +1, and −1. *IEEE Trans. on Circuits and Systems*, CAS-27:451-456, June 1980.

[Cio91] J. Ciofi. *A Multicarrier Primer*. ANSI T1E1.4 Committee Contribution, Boca Raton FL, November 1991.

[Con70] A.C. Constantinides. Spectral transformations for digital filters. *Proc. IEE (London)*, 117:1585–1590, August 1970.

[Cro75] R.E. Crochiere and A.V. Oppenheim. Analysis of linear digital networks. *Proc. IEEE*, 62:581–595, April 1975.

[Gas85] L. Gaszi. Explicit formulas for lattice wave digital filters. *IEEE Trans. on Circuits & Systems*. CAS-32:68–88, January 1985.

[Gra73] A.H. Gray, Jr. and J. D. Markel. Digital lattice and ladder filter synthesis. *IEEE Trans. on Audio and Electroacoustics*. AU-21:491–500, December 1973.

[Her73] O. Herrman, L.R. Rabiner, and D.S.K. Chan. Practical design rules for optimum finite impulse response lowpass digital filters. *Bell System Tech. J..* 52:769-799, 1973.

[Jac70] L.B. Jackson. On the interaction of roundoff noise and dynamic range in digital filters. *Bell System Technical Journal*, 49:159–184, February 1970.

[Jac96] L.B. Jackson. *Digital Filters and Signal Processing*. Kluwer, Boston MA, third edition, 1996.

[Jar88] P. Jarske, Y. Neuvo, and S.K. Mitra. A simple approach to the design of FIR filters with variable characteristics. *Signal Processing*, 14:313–326, 1988.

[Kai74] J.F. Kaiser. Nonrecursive digital filter design using the I_0-sinh window function. *Proc. 1974 IEEE International Symposium on Circuits and Systems.* pages 20-23, San Francisco CA, April 1974.

[Kra94] T.P. Krauss, L. Shure and J.N. Little. *Signal Processing TOOLBOX for use with MATLAB*. The Mathworks, Inc., Natick MA, 1994.

[Lim86] Y.C. Lim. Frequency-response masking approach for the synthesis of sharp linear phase digital filters. *IEEE Trans. on Circuits and Systems*. CAS-33:357-364, April 1986.

[Mar92] A. Mar, editor. *Digital Signal Processing Applications Using the ADSP-2100 Family*. Prentice-Hall, Englewood Cliffs NJ, 1992.

[Mat94] *MATLAB Reference Guide*. The Mathworks, Inc., Natick MA, 1994.

[Mat96] *MATLAB Signal Processing Toolbox User's Guide*. The Mathworks, Inc., Natick MA, 1996.

[Mat97] *Using MATLAB, Version 5*. The Mathworks, Inc., Natick MA, 1997.

[Mit74a] S.K. Mitra and K. Hirano. Digital allpass networks. *IEEE Trans. on Circuits and Systems*, CAS-21:688–700, 1974.

[Mit74b] S.K. Mitra, K. Hirano, and H. Sakaguchi. A simple method of computing the input quantization and the multiplication round-off errors in digital filters. *IEEE Trans. on Acoustics, Speech, and Signal Processing*. ASSP-22:326–329, October 1974.

[Mit77a] S.K. Mitra, K. Mondal, and J. Szczupak. An alternate parallel realization of a digital transfer function. Proc. IEEE (Letters), 65:577–578, April 1977.

[Mit77b] S.K. Mitra and C.S. Burrus. A simple efficient method for the analysis of structures of digital and analog systems. *Archiv für Elektrotechnik und Übertragungstechnik*, 31:33-36, 1977.

[Mit90] S.K. Mitra, Y. Neuvo, and H. Roivainen. Design and implementation of digital filters with variable characteristics. *International Journal on Circuit Theory and Applications*, 18:107-119, 1990.

[Mit98a] S. K. Mitra. *Digital Signal Processing: A Computer-Based Approach*. McGraw-Hill, New York NY, 1998.

[Mit98b] S.K. Mitra and A. Makur. Warped discrete Fourier transform. *Proc. IEEE Workshop on Digital Signal Processing*, Bryce UT, August 1998.

[Neu84] Y. Neuvo, C-Y Dong, and S.K. Mitra. Interpolated finite impulse response filters. *IEEE Trans. on Acoustics, Speech, and Signal Processing*, ASSP-32:563-570, June 1984.

[Neu87] Y. Neuvo, G. Rajan, and S.K. Mitra. Design of narrow-band FIR bandpass digital filters with reduced arithmetic complexity. *IEEE Trans. on Circuits and Systems*, CAS-34:409-419, April 1987.

[Par72] T. W. Parks and J. H. McClellan. Chebyshev approximation for nonrecursive digital filters with linear phase. *IEEE Trans. on Circuit Theory*. CT-19:189-194, 1972.

[Pel80] A. Peled and A. Ruiz. Frequency domain data transmission using reduced computational complexity algorithm. *Proc. IEEE International Conference on Acoustics, Speech and Signal Processing*, Denver, CO, pages 964-967, August 1980.

[She95] O. Shentov, S.K. Mitra, A.N. Hossen, and U. Heute. Subband DFT - Part I: Definition, interpretation and extension. *Signal Processing*, 41:261-277, No. 3, February 1995.

[Vai86] P.P. Vaidyanathan, S.K. Mitra, and Y. Neuvo. A new approach to the realization of low sensitivity IIR digital filters. *IEEE Trans. on Acoustics, Speech, and Signal Processing*, ASSP-34:350–361, April 1986.

[Vai87] P.P. Vaidyanathan and S.K. Mitra. A unified structural interpretation and tutorial review of stability test procedures for linear systems. *Proc. IEEE*, 75:478–497, April 1987.

[Vet88] M. Vetterli. Running FIR and IIR filtering using multirate techniques. *IEEE Trans. on Acoustics, Speech, and Signal Processing*, 36:730-738, May 1988.

Index